NEW DIRECTIONS IN CONSCIOUSNESS STUDIES

New Directions in Consciousness Studies presents some original ideas which will significantly advance scientific understanding of human nature. Written in non-specialised language, the book draws upon concepts and research from history, philosophy, neuroscience and physics to delineate new approaches to the study of consciousness.

Early chapters deal with a range of ideas about our nature, and suggest that the mind can usefully be viewed as a type of dynamic landscape. The account shows how our minds relate to their societies, brains and bodies, and how they differ from computers. Later chapters develop a theory of the basis of consciousness (SoS theory). Using the physical concept of broken symmetry the author shows how the conscious mind may be rooted in temporality; a view that is supported by the occurrence of a wide range of anomalous phenomena. Potentially valuable future lines of research are also identified in this innovative text.

This is a unique and engaging book that will appeal to students and academics in the field of consciousness studies and to anyone who is curious about how consciousness fits into the physical world.

Chris Nunn is a retired Consultant Psychiatrist who used to work in association with the Medical School at Southampton University, UK. He has been Associate Editor of the *Journal of Consciousness Studies* for the past fifteen years.

NEW DIRECTIONS IN CONSCIOUSNESS STUDIES

SoS theory and the nature of time

Chris Nunn

LONDON AND NEW YORK

First published 2016
by Routledge
2 Park Square, Milton Park, Abingdon, Oxon, OX14 4RN

and by Routledge
711 Third Avenue, New York, NY 10017

Routledge is an imprint of the Taylor & Francis Group, an informa business

© 2016 Chris Nunn

The right of Chris Nunn to be identified as the author of this work has been asserted by him in accordance with sections 77 and 78 of the Copyright, Designs and Patents Act 1988.

All rights reserved. No part of this book may be reprinted or reproduced or utilised in any form or by any electronic, mechanical, or other means, now known or hereafter invented, including photocopying and recording, or in any information storage or retrieval system, without permission in writing from the publishers.

Trademark notice: Product or corporate names may be trademarks or registered trademarks, and are used only for identification and explanation without intent to infringe.

British Library Cataloguing in Publication Data
A catalogue record for this book is available from the British Library

Library of Congress Cataloging-in-Publication Data
Nunn, Chris, 1940-
New directions in consciousness studies : SoS theory and the nature of time / Chris Nunn.—1 Edition.
    pages cm
Includes bibliographical references and index.
1. Consciousness. 2. Time perception. I. Title.
BF311.N86 2016
153—dc23   2015017475

ISBN: 978-1-138-92385-0 (hbk)
ISBN: 978-1-138-92388-1 (pbk)
ISBN: 978-1-315-68474-1 (ebk)

Typeset in Bembo
by Keystroke, Station Road, Codsall, Wolverhampton

CONTENTS

Acknowledgements   *ix*
Introduction   *1*

1   Swings and roundabouts   4
A brief survey of the various ideas about human nature that have come and gone since the seventeenth century, ending with a description of principal scientific and other views prevailing at the end of the twentieth century.

2   Picturing 'mind'   18
An account of the usefulness of describing our minds as 'landscapes' in dynamic state spaces that encompass their environments along with brains. Discusses differences between minds and computers.

3   Wetware   29
Describes the neural basis of mental 'landscapes', emphasising the central importance of calcium-ion dynamics in their construction and the essential roles of astrocytes along with neurons.

4   On time   40
*Explores the characteristics of clock time (i.e. the time
of relativity theory and quantum dynamics) and its
differences from experiential time. The chapter ends
with a brief description of 'pre-sponse' findings.*

5   New frontiers   49
*Briefly describes the currently most popular
philosophical, neuroscientific and physics-based lines of
thought about the basis of consciousness. Touches on how
these ideas fit in with a 'landscape' picture of mentality.*

6   Broken symmetry   65
*A proposal that a fundamental neutral monism breaks into
subjective and objective aspects coincidentally with energy
eigenstate measurements. Subjective aspects are thus
founded in temporality and objective ones in spatiality. The
resultant picture of 'subjectivity' is dubbed 'SoS theory'.*

7   Qualia   73
*Briefly describes the concept of 'qualia'. Then discusses
what could differentiate one quale from another.
Concepts of 'qualia spaces', or alternatively
considerations to do with knot theory, offer potentially
useful approaches to the differentiation problem.*

8   Rocks from the sky (part 1)   82
*How best to conceptualise 'reality', followed by
discussion of anomalous events of types that interested
Victorian investigators. Physical anomalies provide
evidence that is probably of greatest relevance to SoS
theory, not least because the theory shows, in principle,
that energy conservation need not always apply in
situations in which the anomalies occur.*

9   Rocks from the sky (part 2)   92
*A discussion of the possible relevance of findings suggestive
of 'reincarnation' and those associated with alleged 'spirit'
communications. Some of the evidence from cases of alleged*

childhood reincarnation and from mediumistic activities is surprisingly robust; possible implications are identified.

10 Life at the edge   99
An account of near-death and end-of-life experiences, emphasising the puzzling nature of the memories associated with them. A discussion of the neural basis of memory and memory-associated phenomena suggests that some of these phenomena are mediated by influences that are independent of clock time.

11 New directions   108
Violations of energy conservation, if conscious-mind associated, would provide evidence to distinguish SoS theory from any conceivable 'quantum consciousness' theory. Suggestions about where to look for them are made. Other especially relevant lines of research include trying to ascertain where the information in near-death experiences (NDEs) comes from and how it gets remembered. There's a possibility that psychedelic-drug experiences might provide a useful substitute for NDEs in enquiries of this sort.

12 Speculations and implications   116
A discussion of implications of SoS theory for understanding social history and the frequent horrors of human behaviour, in so far as it adds further deterministic influences to those with origins in objective aspects of culture.

13 Loose ends and new beginnings   124
Points out that no account has been offered in previous chapters of how any 'back action' of conscious mind on neurology might be mediated. There is, however, a physical idea that might just conceivably provide a basis for explanation. Finally summarises the various experimental approaches needed for further progress.

Appendix: synopsis of the arguments   131

Index   *137*

ACKNOWLEDGEMENTS

Thanks to everyone who has submitted papers to the *Journal of Consciousness Studies* over the past twenty years. Whether subsequently published or remaining unpublished, your enthusiasm and ideas have been inspiring. I'm very grateful also to mentors who have been especially influential in shaping my thinking via online fora and personal contacts. They include Harald Atmanspacher, Erhard Bieberich, Chris Clarke, Jo Edwards, Tal Hendel, Peter Henningsen, Stanley Klein, Alfredo Pereira Jr, Jack Sarfatti and Max Velmans. Many thanks also to the Routledge editorial team, especially Sarah Willis. They suggested detailed amendments that have certainly improved the book in all sorts of ways.

I'd like to remember, too, my first maths teacher who never did manage to convey much in the way of mathematical technique, which I think bored him. What he got across to me instead was the existence and value of broad concepts. He was a roll-your-own, chain-smoking refugee from Vienna who had met Wittgenstein in happier days, but later lost his family to the concentration camps and had somehow ended up in a small school in West Cumbria. He was known to us children only as 'Mr Philip'.

INTRODUCTION

'Know thyself!' is a familiar maxim, reputedly written on the temple of Apollo at Delphi and a joy to pedagogues, if not to their charges, ever since. Often taken to refer to knowledge of one's own individual character, it's also a command to answer general questions about human nature. That's the challenge taken up in this book. What *are* we, really? Opinions have swung wildly over recent centuries, but maybe we shall soon know enough to arrive at some sort of consensus. I want to describe how we can paint the picture of our nature that seems to be emerging from a wide range of sources at present, and tell something of what it should look like when finished. It's a picture that will include and integrate many of the views already held by people the world over, while casting them in new and sometimes surprising lights.

A wise man, who both works as a gardener and is a Glastonbury festival enthusiast, once told me: 'I am a spirit having a human experience, not a human having a spiritual experience.' But maybe he is both. It all depends on what is meant by 'human' and 'spirit'. We can say quite a bit more nowadays about what these words may entail than was possible even twenty years ago. A problem is that our new understandings are *so* new that they haven't yet been properly assimilated into mainstream culture. And, of course, older ideas have an inertia that can impede assimilation, especially when they are articles of faith for some influential group. Nevertheless I hope to show in what follows why any culture wars are unnecessary. It turns out that none of the more coherent views of human nature that still remain influential today are likely to prove wholly wrong; they are probably all partially true in the sense of being applicable in some limited context or from some particular point of view. The new picture, when it is eventually finalised, should automatically show why particular perspectives on our nature have led to apparently irreconcilable conclusions.

The plan is first of all (in Chapter 1) to take a brief look at prevalent opinions about human nature and their historical ups and downs. Then, in subsequent chapters, I'll describe a useful way of envisaging 'mind' that shows how it may relate to our societies, brains and bodies, before getting to mysteries surrounding the nature of consciousness. There is a very real sense in which every one of us *is* our conscious experience. After all our bodies, when unconscious, are effectively nothing more than lumps of meat – albeit lumps with the potential to be *us* once more when we wake up. Questions about conscious experience and approaches to answering them are therefore inevitably central to understanding our basic nature. While we don't yet have firm answers to these questions, we do know quite a lot about what *sorts* of answer may suffice and what lines of evidence will help towards answering them. As I shall describe, we already know enough to sketch the broad outlines of a final picture of what we are and even fill in a few of the likely details. It's a sketch that you may find thought-provoking; at least I hope you will.

When it comes to making progress and breaking the various logjams that impede our understandings at present, the most important practical issues are to do with how best to identify and fill in relevant details of the sketch. What types of research, in other words, are likely to prove most fruitful in helping us to find answers to our questions? I'll be describing a specific theory of the basis of consciousness (dubbed 'SoS' theory for reasons that will be explained later) mainly because it provides useful pointers to potentially valuable lines of research. Readers can make their own judgements about its explanatory value in the light of the relevant, if sometimes controversial, evidence that already exists. Whether or not the theory eventually turns out to be correct in any sense, however, it can nevertheless be regarded as an example of a *class* of theory that is likely to be needed for progress in understanding conscious mind.

One big problem that I faced when it came to writing the book was down to the fact that it deals with topics ranging from fundamental physics, through aspects of psychology to anthropology and history. This is inevitable when enquiring into human nature, which inevitably escapes the confines of any particular specialty. However, I did need to refer to technicalities within a wide range of specialties, especially mathematics, physics and neurology. Should I provide detailed explanations and references for readers who might not be familiar with some particular topic or field? If I tried to do this the book would balloon unmanageably and might become completely unreadable, quite apart from the risk of boring to tears readers who already knew more than me about a subject. Moreover we're all a bit like the Master of Balliol College, who was represented in a piece of undergraduate satire as having claimed that 'what I don't know isn't knowledge'. No superficial explanation of, or reference to, some unfamiliar text is likely to make much impression on anyone when it comes to accepting material that falls outside their existing expertise or experience.

For these reasons, I haven't included the references or 'suggested further reading' lists that are commonly given in books of this sort (apart from a very few exceptions in the case of particularly obscure references, which might be hard to trace). What I've done instead is to make brief mention of technicalities, along with terms and names that can readily be Googled. Wikipedia alone is a great resource and there are lots of other excellent websites that give reliable, up-to-date information about technicalities of all sorts. There's also a lot of propaganda and misinformation out there, of course, but it's usually not too hard to distinguish that from the valid stuff. Academic and professional websites, including Wolfram MathWorld and the like, generally provide vast amounts of reliable information and can give a genuine 'feel' for the reality and scope of just about any subject you care to mention; an appreciation that can be hard to acquire so quickly in any other way.

If you are hazy or curious about any particular name, term or topic that crops up in the book, therefore, please Google it. I know it's a lot to ask, but you'll easily find more and better information than I could provide. And you'll be able to check on whether or not I may have been leading you astray. Any book aiming to tackle the frontiers of our understanding, as this one does, is bound to contain *some* false trails and incorrect facts. I hope there aren't too many, but that's something readers need to assess for themselves. It's worth remembering, though, that received understandings and established 'facts' can also often be wrong. I reckon that around half the 'facts' I was taught at medical school have subsequently turned out to be either incorrect or irrelevant, and roughly the same statistic probably applies to current knowledge in all specialties except mathematics. A lot of what follows is to do with how we might best go about correcting contemporary misconceptions of our nature, along with trying to arrive at a more realistic view of it, so a critical look at all possibly relevant sources of information is needed by everyone interested in the enquiry, not just writers such as me.

1

SWINGS AND ROUNDABOUTS

I'm a far from reliable guide to the souls, spirits, 'atmans', 'kas' and the like that inhabit our bodies according to long-dead priests and philosophers. The ideas involved can seem confusing nowadays, despite the dedication and effort that often went into making them. Even the traditional Christian view that we are beings poised somewhere between the angels and the animals isn't entirely helpful. After all, we may know something about animals but angels are elusive. Luckily René Descartes, in the seventeenth century, put the topic of human nature on a less unmanageable basis than had been the case earlier. He said that we and our worlds divide into the *res extensa* of material aspects, including our bodies and brains, and the *res cogitans* of thoughts, feelings and so forth, which lack material extension. There are at least two major problems with his proposal but it does provide concepts a bit less slippery to deal with than notions of soul or spirit, even though *res cogitans* has a lot in common with 'soul' as conceived by some people.

The first big problem with Descartes' proposal is the difficulty of imagining how something lacking 'extension' could possibly interact with the material world. He himself offered a fudge; he suggested that the two do interact somehow in a special part of the brain (the pineal gland) and nowhere else. Modern neuroscientists should take warning from the fact that evidence apparently supporting his fudge is easy to find. People who are losing their minds, due to Alzheimer's or other, similar illnesses, often have calcified pineal glands and it would have been natural for a seventeenth-century anatomist to conclude that the connection between *cogitans* and *extensa* must be getting blocked by the calcification in these cases. In fact, of course, pineal glands tend to calcify with age and there's no direct correlation with dementia. Nevertheless similar, if less obvious, false inferences quite often appear in the scientific literature nowadays, especially in connection with brain-scan findings – a topic that we'll come to later in the book.

The problem that's less often discussed concerns the fact that our perceptions would seem to belong with *cogitans* but nevertheless *do* appear to have extension. The sun at sunset appears to be out there on the horizon, for instance, and a pain in your toe is perceived as being in your toe, not in some non-spatial entity. Both problems have provided pabulum for philosophers ever since, although they're still arguing and haven't yet been able to settle on any generally agreed solutions. Meanwhile, back in the seventeenth century, theologians continued their discussions of the nature of 'soul' little aware that their worlds of imagination, often shaped by intermittent persecution of one another for heresy, were about to be undermined by Isaac Newton's legacy.

It's ironic perhaps, from his own point of view, that Newton's physics should have had such huge consequences for he himself was at least as interested in alchemy, biblical studies and theology as in gravitation or mechanics. Nevertheless it was only his physics that resonated down the centuries. The motions of the heavens, the tides and the fall of an apple could all be explained, he showed, on the basis of a few simple concepts backed up by some mathematics. The universe, it appeared to many, operates in much the same way as the wonderfully accurate clocks that were being made by Robert Hooke in London and Christiaan Huyghens in The Hague while Newton mused in Cambridge.

The image of a 'clockwork universe' took off in the eighteenth century. Ingenious mechanical automata entertained the aristocracy while intellectuals pursued reason, with its embodiment in logic, mathematics and enquiry into the natural world, to the detriment of faith. The mathematician Pierre-Simon Laplace, an ornament of the Enlightenment, was widely believed to have told Napoleon Bonaparte 'I have no need of that hypothesis' when asked about the role of God in the universe. Souls, as far as the prevailing culture was concerned, were relegated to a conservative clergy or revivalist movements such as Methodism. Inevitably, towards the end of the century, the idea of 'Man a Machine' (the title of a book, *L'Homme Machine*, published in 1747 by the physician Julien de la Mettrie) had become almost commonplace. *Res cogitans* had been subsumed into a clockwork conception of *res extensa*.

As we all know, this approach survived and strengthened over the next two centuries, reinforced by the many medical and other successes that followed from regarding people, or at least their bodies, as nothing more than complex biological machines and then enquiring into their workings. But already in the eighteenth century there were straws in the wind heralding a very different, in some ways more traditional, view of human nature. Count Emanuel Swedenborg was one such straw. He was a Swedish philosopher and mining expert who was subject to frequent 'trances' in the course of which he would encounter 'angels', by which he meant the souls of the dead, and could report back on their doings. His credibility was greatly bolstered by a story (now impossible to confirm or refute) that he had gone into one of his trances while at a high-society dinner in Gothenburg and had described in accurate detail the progress of a fire that was raging through Stockholm that same day.

Anton Mesmer was another such straw in the wind, albeit one who tried unsuccessfully to connect with Enlightenment rationality. He advocated the curative powers of what he called 'animal magnetism', a mysterious and invisible 'fluid' that could be transmitted from therapist to patient by means of special gestures. In a gesture of a different sort towards real magnetism, he also claimed that iron rods were especially potent conveyors of his 'magnetic' influence. Although debunked and dismissed by a range of physicians and proto-scientists, Benjamin Franklin among them, mesmeric therapy proved remarkably popular in pre-Revolutionary Paris whither Mesmer had decamped following marital and other problems in his home city of Vienna, especially to wives of the aristocracy. It attracted followers, some of whom survived the revolution (as did Mesmer himself) who carried it over into the nineteenth century when it diversified and flourished once more, to an extent that has been largely forgotten nowadays.

Mesmerism became a sort of craze for a time in mid-Victorian England, spread by lecturers, demonstrators and out-and-out showmen. Historian Alison Winter described it as follows:

> Mesmeric séances were certainly frequent, even everyday, events. But the Victorians who attended them recorded a fascinating, disturbing, even life-changing experience .... The mesmerist demonstrated the essence of influence; the subject displayed amazing feats of perception and cognition.
>
> (Mesmerized: the Powers of the Mind in Victorian Britain. *Chicago University Press, 1998*)

This sort of thing peaked around 1840 in England. The practical side of mesmerism subsequently dwindled and morphed into hypnotism, while its more theoretical aspects blended with the form of spiritualism that had been heralded by Swedenborg.

In the nineteenth century, therefore, a range of disparate ideas about human nature was pursued with characteristic Victorian vigour and rigour. Luckily the burning of people or books from opposing camps was no longer an option, but careers and reputations remained hostages to the fortunes of debate. Along with the older churches, sects primarily concerned with the salvation of souls diversified and flourished, though they often occupied themselves also with the improvement of bodies. Indeed many well-educated but under-occupied clergymen studied *res extensa*, especially biological aspects, for its own sake as well as in hopes of finding evidence for God's workings and providence. It's probably a good indicator of the spirit of the times that Michael Faraday, the greatest of early nineteenth-century experimental physicists, was a devout Sandemanian – an obscure Protestant sect that had appeared in the previous century.

Advances in biology and medicine, trumpeted by Thomas Huxley, for example, and over-interpreted by the likes of Herbert Spencer (originator of that very misleading catchphrase 'survival of the fittest', which has been a source of all sorts of subsequent mischief), confirmed many in the view that *res cogitans* is entirely secondary to, or a property of, *res extensa*. Like Laplace they 'had no need of that hypothesis' when it came to anything supernatural and tended to regard anyone who disagreed with their materialism as credulous at best and more probably a gullible fool. Later in the century, Pierre Janet, Sigmund Freud and others provided additional grounds for doubting that *cogitans* might be independent of *extensa* with their discovery of unconscious mentality. Hitherto it had been generally assumed that consciousness and *cogitans* were co-extensive but, if unconscious mind belonged with *extensa* as appeared to be the case, why not conscious mind also?

Many dissenters remained, however, who were not directly connected with any traditional religion. Poltergeists, apparitions, hypnosis and phenomena associated with it, plus what we would nowadays call telepathy and placebo effects – all of these seemed to point to the independent existence of a world separate from *res extensa*, albeit one a bit different from Descartes' original concept of *cogitans*. Notions deriving from Buddhism and Hinduism were also beginning to filter through, following the great expansion of Eastern empires. These ideas were often added to the mix. Hybrid sects mushroomed, such as Theosophy or Christian Science.

But spiritualism remained pre-eminent probably because it allied with groups using scientific methodology – statistical surveys, controlled experiments and the like – to investigate strange phenomena. When the Society for Psychical Research (SPR) was established in 1882 most of its original members were avowed spiritualists but it also attracted a range of luminaries from the British academic and social establishments, including six of the country's leading physicists and a future prime minister (Arthur Balfour). Similar organisations, with a similar membership, soon appeared in most of the leading Western nations.

Although people often focus on the nineteenth-century 'wars' between science and religion, individuals of the time were often happy to accommodate both outlooks, as did Charles Darwin himself until his faith was shaken by the early death of a much-loved daughter. 'Man a machine' advocates usually reserved their greatest vitriol for the spiritualists perhaps because they were perceived as perverters of the precious methodology of science. Alfred Russell Wallace, co-conceiver with Darwin of evolutionary theory, got away quite lightly with only a degree of neglect consequent on his interest in spiritualism but the physicist Sir William Crookes, for example, had to endure the equivalent of internet-troll attacks for many years. The most prominent of the physicist/spiritualists, Sir Oliver Lodge, was perhaps lucky in having been able to remain a respected member of the 'establishment', becoming the first Principal of Birmingham University.

The relatively nuanced views of many late nineteenth-century thinkers tended to coarsen and harden among their successors during the first half of the twentieth century. Industrialisation, overpopulation and, above all, war and social upheaval provided a background that constantly encouraged the cynical view of humanity so well portrayed towards the end of the period in L. S. Lowry's horrid (but surprisingly popular) paintings of people as little stick-like figures scurrying through mean industrial landscapes and looking much like ants, those archetypal biological robots. The attitude reached its apogee in behaviourism, which dominated academic psychology for several decades. Behaviourists treated *cogitans* as entirely an appurtenance of a mechanical *extensa* and (consciously!) decided to ignore consciousness altogether. At the same time, people who still retained religious affiliations often veered towards crude forms of evangelism or retreated into medieval understandings.

The fortunes of spiritualism, meanwhile, followed a fluctuating course. It had inherited from mesmerism an unfortunate tendency to harbour fraud, hype and showmanship along with the genuinely puzzling stuff. These always tended to undermine the whole project, especially when they were seized upon by self-appointed defenders of 'proper' science. The huge slaughter of the First World War naturally encouraged many bereaved to turn to mediums in the hope of contacting dead lovers, spouses and relatives. This, in turn, encouraged fraudulence and credulity, resulting in a backlash against the whole field. By mid-century it had become a fringe pursuit with an iffy reputation. Former scientific interest in it had almost entirely switched to parapsychology, which came to occupy a precarious foothold on the fringes of academia. Both the remaining spiritualists and the new parapsychologists had to spend much of their time and energy defending themselves against imputations of fraud or sloppy thinking made by behaviourists and their co-travellers. As a result, the statistical side at least of most parapsychology work is generally more watertight than the methodology described in comparable psychology reports. But that's something that rarely impresses mainstream psychologists.

A whole new slant on the arguments and issues crept up over the course of the century as a consequence of developments in fundamental physics, which cast the nature of *res extensa* in a very different light from that thrown on it by the rigid mechanics of Newton and Laplace. Albert Einstein's relativity theory soon showed that space and time themselves, which are the foundations of *extensa*, are quite unlike the fixed framework that Newton had envisaged. Then quantum mechanics and its successor, quantum field theory, sent Laplace's determinism out of the window at the same time as totally undermining all intuitive views of the nature of matter. It turned out that *extensa* condenses, so to speak, out of fields of potentiality — what actually condenses out on any particular occasion being determined by what sort of 'measurement' is made on a field. Thus, if an apparatus can determine the position of a particle, a particle will appear somewhere. If it can measure spin, say, then a spin in some direction will materialise; the probability of any particular position or spin manifesting being dependent on the rules of quantum theory.

But, of course, the measuring apparatuses, too, are made of the same stuff as what is being measured. Any material 'measurements' must therefore be regarded as reciprocal — what does the measuring is also being measured. This led some to ask whether there is an ultimate measurer; a sort of final court of appeal where the buck finally stops. They concluded that, yes, it's an idea that does make sense and where the buck stops might be when the consciousness of an observer perceives a measurement outcome. Especially popular for a time after the Second World War, this view is less widely held nowadays although some physicists still go along with it, if only because no wholly satisfactory alternative is available. The idea re-introduced a form of *cogitans* into physics and gave it a starring role as perhaps being the principle determinant of any particular *extensa* manifestation. This was an almost complete reversal of the view, prevalent since the Enlightenment and still held by most biologists and psychologists, that *cogitans* is a property of *extensa*.

The end of the twentieth century saw further puzzles arise about the nature of *extensa* and especially its space–time framework. Relativity theory, developed early in the century, is intuitively understandable (sort of!). Physicist John Wheeler neatly summarised it as telling us that 'space tells matter how to move and matter tells space how to curve'. But mysteries associated with quantum entanglement are very far from intuitive. It turns out that any two or more particles that have *ever* interacted, if spared subsequent interactions, must be regarded as forming parts of a *single* whole regardless of any spatial distance that appears to separate them. Moreover it's been shown, in so-called 'delayed choice' experiments, that a later choice of what sort of measurement to make can affect what would appear on any common-sense view to be the *earlier* behaviour of a quantum particle. Neither space nor time can be much like our ordinary perceptions of them, rather as a table may appear solid enough to us but is actually made up of atoms that consist mostly of empty space, held together by fields of force. Various theories (e.g. string theory, loop quantum gravity etc.) attempt to make sense of all this; so far, it has to be said, without much success.

Although the most advanced end-of-twentieth-century physical theories didn't have a definite place for *cogitans*, there were indications that one is needed. For instance, experimental set-ups were devised showing that a single quantum particle can apparently 'know', in some sense or other, about properties of an entire apparatus, including what the outcome of the experiment *would have been* if the particle had behaved differently within the apparatus from the way that it was in fact measured to have behaved. Hard to believe as they may be, outcomes dependent on these might-have-beens were shown to occur in experiments and were dubbed 'quantum counterfactuals', as demonstrated, for example, in set-ups based on the 'Elitzur–Vaidman bomb test' thought experiment.

Behaviour dependent on what might have happened but didn't surely depends on some *cogitans*-like realm because relevant aspects of *extensa* were only potentially present, not actually there and able to influence particle behaviour. In these experiments, as with the view that the consciousness of an observer might be the ultimate measurer, causation appeared to flow from *cogitans* to *extensa*; the reverse of the assumption about direction of causation made hitherto by most 'man a machine' enthusiasts. It's an appearance that could be illusory, however, for it may point instead to an overarching holism between the two Cartesian realms; a unity of a type (similar to, or maybe basically the same as, that manifesting in quantum entanglement) that we don't properly understand.

Before going on to possible new understandings, it's worth pausing to take a look at how the two disparate groups with well-defined views about our nature, the inheritors of the Enlightenment and the spiritualists, have pictured us. Each group, on the one hand, has developed its views on the basis of reason, analogy, argument and evidence, or at least what they took to be evidence. Religionists, on the other hand, claim to base their ideas on revelation handed down through authority. Their views may well have had origins not dissimilar to those leading to the spiritualists' notions, for example, but now obscured by the mists of history. Their opinions and claims have weathered and gathered too much moss over the course of time to be manageable when it comes to an enquiry into the basis of human nature, whatever their value from ethical or moral standpoints.

Those who believe that *res extensa* is all, the convinced materialists, have naturally asked themselves what *sort* of machine man might be like. Descartes himself supposed that our brains may work like the hydraulic machines that embodied the most advanced technology of his day. 'Vital spirits', manifest perhaps in cerebro-spinal fluid and flowing through nerves, might circulate and enable brains to do their work, he wrote. Gottfried Leibniz, a few years later, famously proposed that if we could look inside our brains all we would see would be mechanisms analogous to the gears and pulleys inside a windmill. Both these philosophers, of course, supposed that their mechanisms *couldn't* fully account for *cogitans*. As we have seen, however, this reservation was abandoned in the next century when clockwork automata became all the rage and provided a useful image of our nature according to some thinkers. The pattern of identifying us with advanced technology of the time continued.

Even Sigmund Freud may have been unconsciously influenced to think in this way for we have a lot in common with steam engines according to his view, though he would probably have been horrified had anyone been brave or crude enough to point this out to him. Nevertheless, according to his theory of our nature, a conscious 'ego' is powered by the unconscious 'id' and regulated by the 'superego'. Substitute pistons, boiler and steam governor for his terms and you've got a description of a steam locomotive! Sir Charles Sherrington, the pioneer neurophysiologist, had nicer analogies, comparing our nervous system first to a telephone exchange and then, in a beautiful image, to an 'enchanted loom' in which electric messages shuttle about weaving the tapestry of our everyday experience. The 'tapestry' analogy will be resurrected in future chapters, albeit based on different concepts from those that Sherrington envisaged.

Computers, of course, capped all previous analogies when they came on the scene because they have genuinely mind-like functions. Putting them into automata provided an image of our nature that many found compelling. Stories and films about robots that were more human than humans became almost as clichéd in late twentieth-century popular culture as mesmeric 'passes' and their consequences had been in mid-nineteenth-century media. 'Artificial intelligence' became a major research topic. Although success in realising it has been far more modest to date than was initially hoped, many suppose that artificial minds greatly surpassing our own may be just round the corner. Neuroscientists, realising that our brains don't work like digital computers, have focused on 'neural nets' (simple neural nets can be simulated on digital computers) and have had some success thereby in modelling limited aspects of brain function; enough success to entice them into planning ever more ambitious computer models. Visionaries frequently discuss the possibility of 'downloading' people into computers without any loss of their essential humanity or identity. Mechanical man, reinvented to fit the information age, compelled widespread belief by the twentieth century's end.

The spiritualists, however, had what ought to have been a big advantage. They didn't need to rely on fanciful analogies with machines for their view of human nature because they had first-hand evidence, they supposed, relayed from the departed via mediums. They too had their cultural clichés – varieties of ghosts, ascension into 'the light' and so forth – but they also had a great many detailed descriptions of our true nature to draw on. Of course it wasn't entirely straightforward. 'Spirits' often complained about the difficulty of communication. For example, the deceased 'Frederic Myers' said, according to 'Mrs Holland' (a pseudonym used by Rudyard Kipling's sister, Alice, who was apparently a talented medium), that from 'his' point of view it was like 'standing behind a sheet of frosted glass which blurs sight and deadens sound, dictating feebly to a reluctant and very obtuse secretary'. Mediums themselves often commented on how hard it was to be sure they had distinguished their own thoughts from those of an alleged communicator and to grasp unfamiliar ideas that a communicator might attempt to get across.

Nevertheless a fairly consistent picture emerged from all these alleged communications. According to it, the average person will find themselves, when they have 'died', in an idealised version of the world they left behind; one that appears to them at least as real as the *extensa* they lived in previously. Darker realms exist, but are in some sense 'chosen' by the people who end up there. Efforts are made by more fortunate spirits to persuade lost souls to choose better and move on to happier surroundings. Recent arrivals in the afterlife usually spend their time recuperating from the traumas of earth life, socialising, performing useful tasks and developing their talents. Some eventually choose to re-incarnate while others move on to 'higher' spiritual realms. The re-incarnaters mostly do so in order to learn further moral lessons of a type allegedly available only in our world, while a few incarnate for altruistic reasons – to help fellow spirits who are struggling with the conditions in our world.

It's a picture of our nature and fate that should surely look far more attractive to most people than the nihilism of materialists or the frightening 'judgements' and static, boring eternities advertised by many religions. Why then has it attracted so few adherents? Perhaps a principle reason is that it sounds too good to be true; too much like Father Christmas or a fairy tale with a happy ending and thus is strictly for children, not for grown-ups. Of course this argument can be turned on its head. Maybe we like fairy tales because they reflect what we intuitively understand to be true of ourselves, but any such reversal isn't likely to cut much ice with hard-headed realists.

A better reason for scepticism is down to the first of the main problems with Descartes' *res cogitans*; namely that of envisaging how it could possibly interact with the material world. And it must interact, according to spiritualism, because the evidence on which spiritualists base their claims (some of which is quite good by the way; certainly much better than sceptics allow although how best to interpret it remains a very open question) depends on two-way communication, however imperfect. There's absolutely no possibility of any such communication according to Laplacian physics, which is the variety still assumed to be true by most biologists and nearly all psychologists (presumably because basic school education in physics rarely gets past the nineteenth century). There's no very obvious place for it in modern physics either. Both quantum field theory and the 'standard model' of particle physics, like Laplace, have no need of that hypothesis. Envisaging a basis for spirit communications would have to involve extending present-day physics in some way.

The end of the twentieth century thus harboured contrasting views of our nature, not so much setting science against religion as setting materialism against spiritualism. Science was certainly perceived as endorsing materialism, though its most advanced concepts of what matter is like would have seemed to a nineteenth-century scientist to have little in common with any *res extensa* that he understood ('she' scientists were rare until recently, of course). Moreover what has been called 'promissory materialism' – the claim that, even if you can't explain something now in material terms, it's likely you soon will be able to as science advances – seemed plausible in the light of the huge progress already made, and could apply just as much to understanding human nature as to any other puzzling phenomenon. 'Man a machine' was therefore the culturally respectable, default position. Defenders of it often waxed lyrical about the dangers of a return to superstition and barbarism if ever it were seriously questioned.

All the same, the mainstream version of spiritualism that had descended from nineteenth-century investigators was never entirely extinguished. Offshoots from it, especially parapsychology, kept appearing often to the dismay of self-appointed defenders of scientific orthodoxy. The SPR (the organisation founded in 1882) remains active and many other bodies with similar aims and philosophies have developed since (e.g. the Scientific and Medical Network founded in 1973). Clearly it's a view that assuages some need or intuition held by quite large sections of society; one that isn't adequately met by religion. But how do the views of spiritualists actually differ from those of orthodox scientists? Are they as irreconcilable as appears on the surface? Thinking in terms of *extensa* and *cogitans* can help towards answering these questions.

If 'man a machine', *cogitans* must be either a product or an aspect of *extensa*. For most of the twentieth century, it was assumed to be a product, rather as liquidity is a product of H_2O molecules. Right at the end of the century, though, opinion started to switch to the view that it (or at least the part of it that we refer to as 'consciousness') might be an aspect, not a product, of *extensa*. Technical terms such as 'dual aspect theory' or 'property dualism' were applied to this notion, which was much like a view advocated by Baruch Spinoza in the seventeenth century. Then, in a further development inspired perhaps by both quantum theory and Buddhist philosophies, the idea surfaced that *extensa* and *cogitans* might be equal partners in that *both* are 'aspects'; ones deriving from a split in the foundations of reality whatever these may be. Our everyday world with its *extensa* and *cogitans* aspects, according to this view, depends on a split occurring in an unknown and perhaps unknowable reality that, either potentially or actually, incorporates both aspects.

For spiritualists, however, the *extensa* to which *cogitans* primarily belongs is one that is mostly disconnected from our everyday tangible and visible world; in the popular phrase, it exists 'in a different dimension', albeit one that has to be envisaged as connecting with our 'dimension' in some way. According to both some of the recent orthodox and the spiritualistic views, therefore, there is a split in the reality that underpins our world. The two views differ as to the nature of the split and where it occurs. For orthodoxy, the split is between physical and some (i.e. those associated with consciousness) mental aspects of a single underlying reality; for the spiritualists it involves a separation of physical/mental realities into different realms. Whether these two views are in fact irreconcilable, or whether they are just different ways of describing the same basic phenomenon, is impossible to say. Any attempt to answer this question would depend on knowledge of the reality thought to underlie both *extensa* and *cogitans*. Only mystics claim anything approaching an adequate concept of it, but their reports are too general to be of much help in this connection. String theorists and the like hope to precisely describe fundamental reality one day, but that day has yet to arrive.

Thus the older scientific orthodoxy pictured our nature as shown below:

(*Extensa* → *Cogitans*)
A single reality within which *cogitans* emerges from *extensa*

The newer orthodoxy can be represented as:

(*Extensa* :: *Cogitans*)
A split dividing aspects of an underlying reality. Unconscious mentality would usually be regarded as part of '*extensa*', while '*cogitans*' refers to conscious mind.

The spiritualist concept is less clear, but can probably be represented as:

(**Extensa**/*cogitans* :: *Extensa*/***Cogitans***)
A split in reality dividing aspects that each incorporate both *extensa* and *cogitans*, but in varying proportion as indicated by **bold** type.

The simplest concept is evidently the 'old orthodox' one but, as we'll be seeing in subsequent chapters, it's too simple to be realistic. The 'new orthodox' one is better but unduly vague. It acknowledges that the two aspects are indeed different, as everyday experience confirms, but doesn't tell us where or how the difference originates. In an earlier variant it's about alternative 'takes' on a basically *extensa* world, described as 'objective' and 'subjective' respectively, which is unsatisfactory because it offers no basis for the existence of consciousness. The later variant, envisaging a split in some hypothetical 'fundamental reality', leaves very many questions hanging in the air about both the physics and the psychology of where and how any such split could occur.

The spiritualist concept looks at first sight too messy and complex to merit consideration, with its two separate realms each incorporating bits of the other. However, there's an analogy from physics showing that nature does sometimes work that way. When a split occurs in physics, it's not always a completely clean break but may indeed incorporate bits from each side of a divide in the other, much like the idea of *extensa/cogitans* occurring on either side of a split.

Here's an example: when the universe was very young and very hot, during its first microsecond or so, the electromagnetic force was unified with something called the 'weak nuclear force', according to a forty-year-old theory still thought to be correct. As the universe cooled, the two forces split apart and came to look very different. Electromagnetism is universal and responsible for the structure of almost everything in the world from atoms and molecules, through our brains to power stations. The weak nuclear force is confined to atomic nuclei and responsible mainly for aspects of radioactive decay. Both forces are mediated by particles; photons (which don't themselves carry electric charge) in the case of electromagnetism and three distinct particles in the case of the weak force, two of which do carry electric charge while one (the Z particle) is uncharged. The original force appears to have split apart into two completely distinct entities. Nevertheless any accurate description of the Z particle has to incorporate about 10 per cent photon, while photons turn out to be about 10 per cent Z particle. Clearly, in this context at least, a very fundamental split has indeed left a bit of each side of the divide on the other. Maybe, by analogy, the spiritualist's picture isn't as unnatural as appears at first sight.

While thinking of these various concepts in Descartes' terms of *extensa* versus *cogitans* does help with pinning them down, so to speak, and seeing what they involve, doing so also highlights the two big problems with his idea that were mentioned at the beginning of this chapter. The 'new orthodox' view, on the one hand, tells us virtually nothing about how *extensa* and *cogitans* could interrelate; we know it isn't anything especially to do with the pineal gland, but otherwise we're little further forward. Both the 'old orthodox' and the 'spiritualist' views, on the other hand, imply that 'extension' in some form or other must be a property of *cogitans*, which at least accords with common-sense and everyday experience, though it totally undermines Descartes' rationale for making his distinction. It's time, therefore, to move on from the seventeenth century and see what the twenty-first century has to tell us. We need to think first of all about the nature of 'mind' before getting to what is usually taken to be the closer modern equivalent of *cogitans*, i.e. consciousness.

2
PICTURING 'MIND'

'She knows her own mind' people often say of someone, whether in admiration or exasperation. But does she actually know it, or is she just unusually firm in expressing the opinions it throws up for her? The problem is that a lot of her mind isn't 'her own' in the first place. It's something that she shares widely with others, because much of it originates in her genetics on the one hand or in the society surrounding her on the other. Only a relatively small part of her mind has purely personal origins, despite what she often thinks. There follows some anecdotes to show people's minds behaving in ways that aren't up to them as individuals. They tell of just a few bizarre historical happenings although similar events are occurring all the time, all over the world.

Back in 1844, the Mormons faced a crisis. Joseph Smith (their founding prophet), handsome, tall, elegant and euphonious (these details matter as we shall see), had been shot dead by a mob of unbelievers. Most of the faithful had come together in their home town of Nauvoo, Illinois to try to decide on their future leadership. There was disagreement. The senior surviving member of their 'council of apostles' (Sidney Rigdon) argued that Smith was irreplaceable but that he, Rigdon, should nevertheless be nominated 'protector of the church' – a title harking back in its implied ambition, perhaps, to Oliver Cromwell's 'Lord Protector'. Then another apostle, Brigham Young, got up to speak. He was stockily built, no oil painting and with a far from musical voice. Nevertheless a significant proportion, perhaps a majority, of his audience saw him transmogrify into the beautiful form of Joseph Smith and heard him utter words in Smith's melodious voice – which effectively both undermined Rigdon's claim about Smith's irreplaceability and left no doubt as to who should assume the mantle.

Not so long ago, around the late 1980s, as many as four million Americans plus people in other countries were claiming to remember having met aliens in alarming encounters. Most aliens were said to be three or four feet high with greyish skin, large tear-drop heads, no ears or hair, big dark eyes and lipless mouths. They would somehow zap their victim, who was often a woman and frequently in bed either alone or while her partner slept. The next thing she would recall was waking up in some sort of medical facility in a spaceship. She would be examined and probed, most often gynaecologically. Samples were often taken from her ovaries or elsewhere. Most times, the process was painful. Gadgets of unknown purpose might be implanted in her nose or head. Then she might have a telepathic conversation with one of the aliens. Given the circumstances, these conversations were almost always remarkably banal. Next thing she would find herself back in her own bed, feeling sore and upset and maybe with a nosebleed or a few strange marks on her skin to show. When men were involved, the procedure was much the same, except that the samples were often of forcibly extracted sperm. Naturally there were big meetings and conferences about all this. The stories told by abductees became ever more bizarre over time; the variety of aliens and spaceships proliferated; they took to often floating people around on beams of blue light and 'independent' witnesses claimed to have seen such marvels. Star abductees became involved in what could look to the uncharitable like a competition to see who could be abducted most often. One or two got away each night for anything up to a week or so. Some were even snatched from the venues of the very conferences they were attending. It all peaked and then fizzled out. Abduction was being reported only rarely by the turn of the century, usually by people living in cultural backwaters – with apologies for the 'backwater' epithet to a town in central Scotland that harboured several new abductees as late as the year 2001!

My final anecdote is to do with a mother of two who went to Paris with a rich boyfriend. The chauffeur-driven car in which they were travelling after leaving their hotel hit a wall and, sadly, she was killed. It was the sort of event that would normally elicit no more than dismissive comments from anyone who heard about it and had not known her personally – 'serve her right', 'she had it coming', 'how sad', 'poor thing' – that sort of remark, dependent on the hearer's degree of charity.

Yet, when her funeral took place on 6 September 1997, up to half of the adult population of the UK, perhaps 20 million people, felt strong and perfectly genuine grief. For the woman was Diana, 'the people's princess'.

Whatever 'mind' may be, these stories show that it can certainly work in strange ways, throwing up experiences that have little or nothing directly to do with our own individual bodies and brains, and that may not even correspond to any real event. It's pretty certain Brigham Young didn't actually turn into Joseph Smith for however long his speech took, and few would agree that aliens might really have acted in the way that people 'remembered'. What would be the point, after all, of temporarily snatching more than four million people? How much time and how many spaceships would you need to process them all? All the same, mind must usually connect somehow with a real world out there and produce sensible outcomes – we would have become extinct long ago if it didn't.

The stories actually tell us quite a lot, in a general sort of way, about how minds do work. The Mormons' shared concerns and experiences got together within their individual brains and somehow threw up the same percept; one that effectively solved their leadership problem for them. That's something that goes on every day for everyone when they focus on a problem and then 'see' a solution to it. What happened with the Mormons was unusual mainly because so many of them shared the same intense focus and all quite literally saw the solution simultaneously, whereas everyday concerns usually take a unique form for each individual and are generally less intense, with any 'seeing' being metaphorical. The lesson here is that 'mind' has to be viewed as extending beyond individual brains occasionally condensing, so to speak, into a group mind.

The 'alien abduction' experience offers a slightly different lesson because it went on for two decades or more and affected many widely separated people. They weren't in any sense 'mad', by the way; studies showed them to be normal people from a normal range of backgrounds who were perhaps slightly more imaginative than the average person but that was the worst (or best, depending on your point of view) that could be said of them. The experience was based on a whole lot of ideas lodged at the back of peoples' minds, ranging from folk tales about abduction by fairies, through science-fiction stories to apparently credible reports of UFO sightings. The 'glue' that enabled the whole thing to take off and affect so many was provided by a number of well-meaning believers in abduction who used hypnotism to 'recover' memories of it from others. They supposed that vague, unexplained physical symptoms or emotional upsets, plus an inability to recall some period of time, might be down to abduction and traumatic but repressed memories of it. Bringing the memories to light via hypnosis would be helpful, they thought. Although there was little awareness of false memory syndrome at that time the would-be do-gooders really ought to have been aware of how easy it is to implant false ideas in people via hypnosis, but apparently they ignored this and got ever more excited about the 'evidence' they uncovered. They are the ones who seem to have honed both the image of little grey, hairless people with big black eyes that became so widespread, plus the ideas about what went on in the spaceships.

The Diana story shows that it's not only percepts and ideas that can have a basis in shared mentality, but also strong emotions. The tabloid press had built an image of her over the years as an archetypal, wronged and fragile, doomed beauty, which may or may not have had much connection with her real character. Nevertheless many felt that they knew her personally and each reacted to her death as if the media-constructed image had been an actual close friend, relation or lover.

There's a big question to do with what sort of overall picture could possibly portray both mind's evident dependence on our shared cultural, social and physical environments and the fact that it is orchestrated, so to speak, within individual brains. The first step needed towards finding an answer is to recognise that mind isn't so much a static thing as an ongoing process.

All sorts of inputs contribute to mind, which are taken up and processed by brains whose outputs eventually feed back in one way or another to produce new inputs. The Behaviourists' whole approach to psychology, which was dominant for much of the mid-twentieth century, depended on isolating this first step. They tried to analyse all of psychology in terms of inputs to a 'black box' (i.e. the brain) and subsequent outputs from it. Where they mainly went wrong was in trying too hard to reduce everything to simple components and in not seeing that their methods were never going to tell them all that much in any case. In fact mind is a holistic process where the whole is greater than a sum of the parts, so you're bound to go astray if you try to reduce it to separate bits. It is actually like the swirling patterns in a waterfall. Try to pick out individual scoops of water and the pattern changes or disappears – and caging the swirls was in effect what Behaviourists aimed to do, though naturally with very little success.

The next step therefore involves finding a representation of mind that shows its holism. And, because it's a dynamic process arising from a great many sources, there's an obvious type of representation that boils down to nothing more than an extremely complicated graph. If you wanted to picture the height of a ball above the ground after it is thrown, you'd plot a graph with height above ground on the 'y' axis and time on the 'x' axis, and you'd get a parabolic curve shown in two dimensions. If you want to plot a complicated dynamic process in the same sort of way, you need axes for each element contributing to the process. In fact you need six axes for each element: three for its position at any particular moment and another three for its direction of motion in each of the three spatial dimensions. Instead of a graph in two dimensions, you get a representation in vast numbers of imaginary dimensions when it comes to any really complex process. To represent mind you'd need many trillions of imaginary dimensions. This isn't such an absurd idea as it may seem at first sight because quantum theorists have to use a similar notion every day. They picture their 'wave functions' as embedded in something called 'Hilbert space', which has an *infinite* number of (strictly notional, of course) dimensions, not just the many trillions needed for our purpose here.

Of course there's no hope of ever being able to actually construct this sort of abstract picture of mind from information about the dynamics involved; the number of variables is too vast and the uncertainties about their details too great. Nevertheless, if it were possible to paint it, the picture would have interesting general properties. For instance, it would look like an incredibly elaborate and baroque mountain landscape with all sorts of tunnels and aerial walkways added on, albeit one seen on a geological timescale so that it was constantly heaving and changing. And this analogy can be taken quite a bit further because the dynamic events occurring at any particular moment can be regarded as 'raindrops' falling on the landscape and then being channelled down some particular 'valley'; the choice of 'valley' depending on where each drop happened to fall. In other words, the outcome of any particular event is determined by the 'contours' of the overall landscape.

Unlike an actual mountain landscape, which is always three dimensional, our imaginary one 'exists' in a vast, but also variable, number of dimensions. This variation happens because the dynamic events from which it is constructed change from time to time, often from moment to moment. Your dynamical circumstances when you are diving into a swimming pool are very different from when you are resting in bed, for example. When the dimensionality of the 'landscape' changes, its configuration is also likely to change; some 'valleys' will be obliterated and new ones may appear. Thus any particular type of 'raindrop' may have widely different fates at different times, depending on the type of 'landscape' existing when it happens to fall. The drop might land on a mountain top, rush down a river and contribute to powering a hydroelectric station, say, or it might plonk into a marsh and get swallowed by a frog.

The 'raindrops' have all sorts of origins. Some are down to happenings in our surroundings that are picked up by our sensory organs – events that we register as sights, sounds or smells, for example. Others originate in our bodies, such as the information from muscles that tells us about our movements. Most (around 90 per cent or more would be a reasonable guesstimate) are due to internal goings-on within the brain. But what are the 'valleys' in the landscapes of mind that are watered by these raindrops? Technically they're called 'attractors'. Actually (if that's the right word to apply to what are basically mathematical concepts!) they are memories, or at least representations of memories. Some of these memories, often those having the widest and deepest valleys in the landscape, are down to our genetic inheritance – our DNA. Others hold whatever we have learned and remembered over the course of our lives. Some of these non-genetic memories are strictly personal, but most of them originated in the families, societies and cultures that surrounded us all our lives. The associated 'valleys' naturally have to be regarded as co-extensive with the sources of the memories that they represent. They spread beyond the brain into associated environmental dynamics, thus showing *how* our minds can sometimes appear to belong more to our environments than to our personal selves, in ways illustrated by the stories told earlier.

A big plus of this picture is that it shows, in a general sort of way, how our minds actually work. Contrary to popular belief, they don't have much in common with any contemporary computer. Information theorists have often been mystified as to how we can do things like recognising a face from any angle (except upside down) and in all sorts of lighting conditions, quicker and often more accurately than computers can manage. After all the processors in computers work millions of times faster than the nerve cells in our brains so computers ought to find recognition tasks a doddle, one might suppose. What stymies them is that they have to do everything in sequential steps. They, on the one hand, have to scan a little bit of face and then compare it with a stored memory, then another bit and another comparison, and so on for many, many steps, until finally they can say whether they've got a match between input and memory. We, on the other hand, do it all in just a few steps, some of which are set up before we even start looking. If you are meeting someone you know at an arrivals gate in an airport, for example, your mental landscape will already contain a 'valley' corresponding to your friend's size, shape, posture, manner of walking, facial appearance and so forth. As soon as the appropriate informational 'rain' reaches your eyes, even if it's only the equivalent of a light shower, it'll sluice down the appropriate 'valley' and – bingo! – there's your friend. If he or she was unlucky enough to be, in that rather sinister official phrase 'a person of interest', you've very probably beaten the facial recognition software working off the security cameras.

The ability of 'landscapes' to reconfigure themselves according to dynamic circumstances is also essential to using tools. When we get in a car to drive, for example, a whole landscape learned in the course of driving lessons snaps into existence. It's a landscape that includes dimensions relating to aspects of the car's dynamics, what happens when the steering wheel is turned and so forth, along with dimensions dependent on the behaviour of our bodies and brains. That's why a car can feel like an extension of themselves to experienced drivers (and presumably why damaging the car can be such a painful experience!). The car, or at least dimensions deriving from its dynamics, is *part* of the landscape of its driver's mind. As soon as you get to your destination and get out of the car some whole new landscape pops up and you are all set to cope with a new set of demands.

One downside of our way of doing things is that, if you have the wrong type of landscape set up in your mind, you may not notice the obvious; the 'rain' relating to something right in front of your nose doesn't land near any valley, so to speak. It's called 'change blindness'. In one famous experiment, for example, people were asked to concentrate on details of a sporting event (counting the number of passes between basketball players), thus setting up a mainly 'watch the balls and do the arithmetic' sort of landscape. While they were so occupied, a man dressed up in a gorilla suit wandered across the scene. Half of the people never noticed him!

Another similar downside has resulted, in all likelihood, in our need to spend a lot of our lives asleep. Of course it's often advantageous to animals like us to be tucked up somewhere safe and husbanding our energy resources, but nevertheless all animals must sleep even when it's dangerous for them and they'd apparently be better off doing something else. Herbivores on the African plains, for instance, at risk day and night from predators, tend to sleep in short snatches while others in their herds remain awake and on watch; dolphins, who might drown if they slept like us, go to sleep with one half of their brain at a time, leaving the other half to keep them swimming. Sleep is clearly essential for some reason, and the main reason probably lies in the ever-increasing 'rigidity' of mental landscapes with use. The more often you perform a familiar task, the deeper the relevant 'valleys' become. It's as if the 'rain' tends to erode them and transform them into permanent landscape features. That's a very bad thing because it means that it becomes ever more difficult to notice anything new that may be happening around you and to react appropriately. Sleep, which imposes an entirely different dynamic on the mind from its waking dynamics, prevents this sort of development from getting out of hand.

The most consistent outcome of sleep deprivation experiments in humans is reduced cognitive flexibility (i.e. reduced ability to switch tracks and take new information on board), and in animals is reduced behavioural flexibility, which is entirely consistent with supposing that the main function of sleep is to restore landscape changeability. If sleep deprivation was too prolonged (as happened in some earlier animal experiments that, happily, would now be regarded as unethical), animals often died. These experiments were not only unethical but also pointless because the deaths may well have been due to the stress of measures needed to keep animals awake continuously and for very long periods of time, rather than to lack of sleep per se. I'll spare you the details of how it was done, but it involved conspicuous cruelty.

Prolonging sleep deprivation experiments didn't tell us anything additional, therefore, about consequences of ever-deepening landscape valleys. There is good evidence, however, that sleep helps with memory consolidation, especially when it comes to learning new skills. If people practise some new task, for instance, they tend to be better at it after a period of sleep than after a similar amount of time spent doing something else. New memories are, of course, equivalent to forming new valleys and memory consolidation can be viewed as the process of establishing the potential for a new valley to manifest in the 'right' dynamic circumstances. Sleep may well have a secondary function, therefore, of allowing rehearsal of new dynamics, which presumably can't be undertaken as readily when there's a lot of old landscape activity going on as is the case during wakefulness.

Given the problems with our sort of 'dynamic landscape' mentality – i.e. its propensity to turn fictions into perceived realities, which worked for the Mormons but is usually disadvantageous, its liability to ignore surprising happenings when the 'wrong' landscape is present and its need to go 'offline' for long periods – what are its main upsides? An obvious, specific answer lies in its ability to almost instantly reconfigure itself, whenever the occasion demands, to perform previously learned skills of all sorts. Perhaps the most important general answer to the question, though, is that our landscapes automatically attach *meaning* to information. The fact that the added meanings can sometimes be spurious doesn't outweigh the advantages of automatic attachment, though all of educational and most academic activities can be viewed as endeavours aimed at tipping the balance further in the 'advantage' direction.

It's an answer, however, that needs quite a bit of unpacking. We often suppose that information and meaning mean the same thing, but they don't. The father of information theory, Claude Shannon, specified that our 'bits' and 'bytes' must be treated as meaning-free. Perhaps the best concept of information for everyday use, if you're not a computer scientist or a telephone engineer, was proposed by Gregory Bateson, a twentieth-century anthropologist and polymath. He said it's 'a difference that makes a difference'. That, too, is inherently meaning-free. After all, when a pebble rolls down a hill and hits a stone, information about its mass and speed is transmitted to the stone but the information doesn't *mean* anything to either of them. Neither would be bothered if the information had been a bit different, and the only observable difference in outcomes from differing 'information' would most often be no more than a small variation in the amount of waste heat resulting from the collision.

The same applies to the information in computers. The information as such is inherently meaning-free and usually only gets meaning from *us* via our input into hardware and software design. People often forget this and are then surprised by how hard it is to get computers to find their own meanings, as is needed for any truly independent artificial intelligence. There are a few programs able to generate their own meanings and attach them to incoming information (e.g. Melanie Mitchell's 'Copycat') but it's fair to say that they are clunky at best and poor imitations of our automatic integration of the two.

The problem computers have is that, the more information they can access for processing, the more possible combinations of it there are. And the number of combinations goes up *exponentially* as the amount of available information increases. This is known as the 'combinatorial problem'. Any sophisticated form of intelligence needs access to lots of information. Unless there is some shortcut way of sorting out the particular bits that are likely to be needed for whatever it's trying to do, the would-be artificial intelligence is going to take years going through all the possible combinations. Even the most speedy of supercomputers won't be much better off than a basic laptop because of the exponential, not linear, increase in number of combinations with amount of information available. It has sometimes been suggested that massive parallelism might solve the 'combinatorial problem' limitation but, like increasing speed, finite parallelism can produce only linear increments of computational power, not the exponential increases needed to resolve the problem or the shortcuts needed to evade it.

Quantum computers, however, with their potential for *infinite* parallelism, might conceivably do better if and when they appear. There has been a range of speculations (that I'll describe in Chapter 5) about the possibility that our brains might *be* quantum computers. One good reason for scepticism is that quantum computation is expected to be particularly good at doing things that we are very bad at, such as finding the prime factors of numbers. Another is that there would be no reason to expect a quantum computer to have the limitations or 'downsides' to its 'mind' that were mentioned earlier. A third is that the 'shortcut' allowing us to evade the combinatorial problem both entails these 'downsides' and has nothing to do with quantum computation.

The shortcut that we have is a property of our landscapes, which are built from memories and rapidly adapt their form to suit whatever dynamic circumstances we find ourselves in. There are valleys that ensure that when a smell of cooking reaches our noses, for example, the information (the 'difference that makes a difference') that it carries will cascade down to trigger all sorts of pre-determined secondary effects, while a smell of petrol will head off down a different valley, triggering different effects. The largest and deepest valleys derive from genetic memories, but they have innumerable side gullies formed as we learn to navigate our physical and social environments during the course of our lives. And information that doesn't meet up with some appropriate 'valley' usually goes nowhere. Digital computers, in contrast, have to search the whole of what are in effect planar memories in order to determine what to do with incoming information; there are no shortcuts for them. Artificial neural nets do embody 'landscapes' analogous to ours, but they stand in the same relation to ours as does a jelly mould to the Himalayas and there's no practical way of much increasing their complexity.

What this chapter shows is that our minds, far from being 'islands complete unto themselves' as we often tend to think, actually do their thing by extending deep into genetic and cultural history and far into physical and social environments. The 'embodied mind' movement of the late twentieth century recognised this to some extent; Francisco Varela, building on the work of philosopher Maurice Merleau-Ponty, was a moving spirit behind it and his work has been continued by Evan Thompson and others.

In a way, though, 'embodied mind' was not radical enough. Philosophers have since developed a basically similar approach, the 'extended mind' hypothesis, but even the most radical of them (e.g. Andy Clark) don't always seem to appreciate the depth of mind's roots in biological and social history, or the degree to which its branches spread into its surroundings. Nor do they always recognise the connections between embodied and extended mind because separate authors wrote separately about these topics using rather different vocabularies. In fact, of course, there is no 'in principle' difference between pinching yourself to remember a date or writing it in a diary; equally, your fingernails are just as dead as a table knife and you could use either to scrape a spot of dried food off a plate. As all DIY enthusiasts and other tool users know, or soon learn to know, bodies extend into their environments with no fixed boundaries between the two. Our notional 'landscapes of the mind' are vaster, more elaborate and more extensive than we readily imagine, which leads us straight on to the next topic. How do our brains manage to encompass and embody the enormous complexity of the dynamics that they harbour?

3
WETWARE

For all of its billions of neurons, space inside our brains is limited. Mental landscapes, with their many trillions of notional dimensions, have to be compacted down, so to speak, in order to fit within a brain's three real spatial dimensions plus its one dimension of time. Of course the actual mapping is the other way round; the landscapes are no more than imagined representations of real dynamical events within the brain. As we saw in the last chapter, however, thinking of the dynamics in 'landscape' terms allowed us to see how our minds extend seamlessly from brains into physical and social environments, as well as enabling us to understand otherwise puzzling aspects of their limitations and strengths.

Imagining how such a landscape would have to look if reduced to three dimensions is also instructive. To get a feel for what's needed, it could be worth searching for 'tesseract' on Wikipedia. Tesseracts are four-dimensional cubes and the Wikipedia entry has representations of them rotating in three dimensions. Cubes are simple shapes, of course, and the models shown are only one dimension down from an 'actual' tesseract. Nevertheless they look quite complicated. Representing mental 'landscapes' in 3D requires the reduction of very complex 'landforms' (most of them 'strange attractors', which can take all sorts of weird or beautiful shapes) by astronomical numbers of dimensions. The only sorts of 3D structure that could provide adequate representations are fractal ones, such as the 'Mandelbrot sets' that adorned so many T-shirts twenty years ago but with the difference that they both *are* 3D, not 2D as on the T-shirts, and are constantly morphing into new fractal patterns with the morphing process itself being fractally structured in time, perhaps, albeit probably over a narrower range of scales than are involved in the spatial patterning.

This at once rules out any possibility that action potentials or neurotransmitter release in synapses could be at the basis of mentality. Any fractal-like structure shown by these phenomena is limited to a small range of spatial scales and is often based on relatively fixed anatomical features. They couldn't possibly represent the universality and fluidity required of anything able to embody our 'landscapes'. What they actually do, according to this metaphor, is provide the informational 'rain' that falls onto landscapes of the mind, is channelled down valleys and may, over the course of time, create new landforms. Nerve impulses (action potentials) carry many of the signals belonging to Sherrington's 'enchanted loom' (see Chapter 1) but they don't design the pattern that it weaves.

Electromagnetic fields in the brain are much more promising candidates for the role of pattern designers. Several theorists (e.g. Johnjoe McFadden, Sue Pockett, Herms Romijn) have suggested that they *are* (conscious) mind. Moreover classic experiments by Walter Freeman and colleagues showed that these fields can and do form the type of fluid 'landscape' envisaged in this book – or at least these authors showed that relatively simple, dynamic, electromagnetic 'landscapes', ones that reflect perceptual activity with variations related to the context in which it occurs, can be visualised in the olfactory lobes of rabbits and in parts of the brains of salamanders. These systems are relatively simple and can be investigated using available technology. It hasn't been possible to extend empirical findings of this sort more widely because the complexities of most perceptual and cognitive neural systems defeat current experimental techniques.

Electrical fields in the brain depend on ion flows, sometimes passive and sometimes actively pumped, through dedicated channels in cell membranes along with the less selective 'gap junctions'. The latter, when open, allow direct electrical and ionic contact between the interiors of cells that they connect. The flows and the fields are, in a sense, aspects of a single process since the fields that ions produce have secondary consequences for ion flows, both directly and mediated by a range of cellular mechanisms. Nevertheless it seems more sensible to look for the basis of 'mind' in general, leaving aside questions of consciousness for the time being, in the flows themselves rather than in the fields that they generate. Of course the whole situation is impossibly complicated as individual ions (of which the most important are sodium, potassium, chloride, calcium and magnesium) influence one another's flows and are driven or regulated by some fifty or more neurotransmitters and neuromodulators, many of which have more than one type of receptor, each type producing different effects on ion flows.

Despite the complexity and the difficulty of teasing out particular physiological contributions to the basis of mind, there is one ion that seems to play an especially central role. This is the calcium ion, which is known to have a variety of essential parts to play in both neurotransmission and in early stages of memory processes. It's this dual role that makes calcium especially interesting from our 'landscapes of the mind' point of view since these landscapes are built of the neural dynamics and formed by the memories that calcium influences in so many ways. $Ca++$ dynamics are unlikely to tell the whole story about the brain's instantiation of 'mind', but they may tell a significant part of it. And they have the additional advantage that they are more accessible to experimental visualisation than the dynamics of other ions. Calcium-ion behaviour is probably the single best token indicator that we have of the physical dynamics of mind.

The behaviour is fractal or pseudo-fractal (i.e. may possess different fractal dimensions on different scales). Pseudo-fractality would actually better represent our 'landscapes' than the more ordered, true fractality since the landscapes themselves are likely to be pseudo-fractal. At the smallest scale, tiny packets of ions are released within the smallest branchings of nerve cells (i.e. dendritic spines) as neurotransmitters are released at synapses; most of these ions are quickly bound to proteins but some diffuse for short distances. Rates of diffusion of ions from spines into dendrites are dependent on the shape of the spines, which is constantly changing (what controls these shape changes isn't known, though it is known that dendritic spines wave about and lengthen and contract almost like waterweed in a strong current). On slightly larger scales there is flow through calcium channels into dendrites and nerve-cell bodies, regulated by a whole range of factors including electric fields (there are dedicated calcium channels in nerve-cell membranes whose opening depends on the magnitude of the trans-membrane potential: so-called 'voltage-gated' channels).

Just recently techniques have been developed for looking at calcium dynamics in animals that are awake (previous techniques were confined to bits of brain grown in the laboratory or to anaesthetised animals). They're difficult to do and only smallish brain areas can be looked at. Nevertheless a study carried out on mice, for example, found that patterns of calcium release within dendritic arbours belonging to a part of the brain (the hippocampus) reflected a mouse's awareness of its position in a virtual maze. If it ever gets easier to do this sort of study, the chances are it's going to tell us a lot more about the detail of neural events underpinning mind than all the more standard techniques combined (i.e. electroencephalogram (EEG), functional magnetic resonance imaging (fMRI), positron emission tomography (PET) and single-cell recordings).

One particularly important concomitant of these fields of varying Ca++ concentration is their effect on a group of similar proteins that are widespread in the brain and elsewhere, referred to as CaMKII (calcium/calmodulin dependent protein kinase II). These proteins are activated by increasing Ca++ concentration. The amount of time for which they stay active is proportional to concentration until a threshold is reached when they switch permanently into their active form. When active, CaMKII has a whole range of effects. An especially important one in the brain is to help with the formation of long-term memories by aiding the development of 'long-term potentiation' in synapses. Long-term potentiation results in a synapse becoming permanently more likely to release its neurotransmitters when a nerve impulse arrives. Thus our fractal patterns of Ca++ concentration have direct consequences for a protein that not only provides a short-term memory, via its alteration to a temporarily activated state, but may itself switch to a more permanent memory if the Ca++ concentration is high enough, while at the same time promoting the development of other permanent memories. Since patterns of CaMKII activation must mirror calcium-ion behaviour, we have a memory mechanism well able to create the 'valleys' needed for our landscapes of the mind.

Larger scales of fractal patterning are needed to complete the picture. At the very largest scales brain-wide activation patterns, such as are shown to occur by EEGs or fMRIs for instance, must have consequences for overall Ca++ behaviour. Aspects of EEGs in particular are often fractally structured, both spatially and temporally but we seem to be missing the intermediate scale that is needed – a scale to bridge the gap between goings-on at the various levels *within* neurons and those that occur in extensive brain areas. Or at least the intermediate scale *was* missing until quite recently for it depends on new appreciations of the importance of astrocytes. It has been known for about fifteen years that sheets of astrocytes grown in the laboratory can harbour waves of varying Ca++ concentration, spreading through a sheet without apparently being impeded by cell boundaries. That's just the sort of behaviour needed to supply our intermediate fractal scale, and it has been found to occur *in vivo* as well as *in vitro*. Astrocytes do support a trans-cellular Ca++ dynamic.

Since astrocytes appear to supply an essential component of the brain's representation of 'mind', it follows that they must play essential parts in cognition and the like. Until quite recently it had been supposed that their role involved nothing more than support of neurons' metabolism and blood supply. They were thought to be no more than the delivery drivers, cooks and cleaners that kept neurons going. No one supposed that they might themselves do significant information processing, let alone sometimes *control* the behaviour of neurons. Yet this is the picture that has emerged over the last few years. They do indeed have essential parts to play in mental functioning.

As their name implies, astrocytes are star-shaped cells with many branches, up to a thousand or more each, which are generally more symmetrically disposed than the otherwise not dissimilar 'dendrites' of neurons. Dendrites belonging to different neurons are often linked by gap junctions, as are astrocytes, and it appears that these junctions also sometimes connect dendrites directly to astrocytes, thus allowing direct electrical and ionic connectivity between the interiors of these two different types of cell. Astrocytes lack, however, the long axons or 'nerve fibres' that neurons possess and, although they have receptors for a range of neurotransmitters and neuromodulators, they don't form synapses with one another as do neurons. Nevertheless it seems that they participate in neuronal synapses and can regulate synaptic activity to an extent. The concept of a 'tripartite synapse' involving transmitting and receiving neurons along with an astrocytic contribution is becoming ever more popular in the specialist literature.

Early estimates of their numbers sometimes suggested that there are as many as ten astrocytes for every neuron (a figure that can still be found in some textbooks). More recent estimates put the ratio at probably somewhere between 1.5:1 and 1:1 in humans. Total numbers of cells in the brain, and even their ratios, are notoriously hard to estimate with any accuracy. Estimates made of the total number of neurons in the brain have varied by as much as an order of magnitude over the course of time, and astrocytes have been far less studied. What *is* entirely certain is that the astrocyte/neuron ratio increases as one goes up the phylogenetic scale. The flatworm *c.elegans*, for example, is known to have a ratio of only 1:5. Even rats probably have a smaller ratio than ourselves; moreover it has been claimed that human astrocytes tend to be larger and with more branchings than those of rats. Individual astrocytes cover neuronal 'domains', each of which, in the case of rats, is said to include around 90,000 synapses. In our case, however, the estimate is that a 'domain' can cover around two million of our synapses, twenty times as many as in rodents. At one time it used to be said that the main or only factor distinguishing our brains from those of other animals is their greater size relative to body weight. It wasn't easy to account for our far greater cognitive abilities on this sort of basis. But it's now looking as though our really important advantage over other animals could lie in the number, size and complexity of our astrocytes.

New discoveries about reciprocal relationships between astrocytes and neurons are being reported every few weeks and the picture, like nearly everything else to do with the brain, is looking inordinately complex. At present the details are of mainly specialist interest and in any case are likely to be revised quite often before any canonical version emerges. Moreover, while astrocytes specialise in relating to and modulating the activity of dendrites and synapses, other cells called oligodendrocytes relate mainly to nerve axons and appear to have equally important controlling roles over them. Douglas Fields, a neurophysiologist, referred to glial cells (i.e. astrocytes, oligodendrocytes and microglia) as 'the other brain' in the title of his excellent book, now a little outdated as it was published in 2009. Nevertheless it remains true that, in the ballroom of the brain, neurons may be the dancers weaving informational patterns but glial cells, especially astroglia, can be thought of as an orchestra shaping the dance and possibly contributing some of its choreography too.

It's clear that there is something of a revolution brewing in neurology and neuroscience as a consequence of these discoveries. What's especially relevant from the point of view of the model of mind offered in this book is that it was a *predicted* revolution. Predictions of new phenomena are fairly common in physics but very rare in neuroscience where progress normally depends on developing some new investigatory technique (e.g. fMRI, PET, EEG, single-cell recordings etc.) or looking at outcomes of 'natural experiments' (e.g. head injuries) and then developing theories or concepts later on to account for any new phenomena that are observed.

However, it was already clear more than ten years ago that considerations to do with higher mental functions and calcium-ion behaviour (i.e. a crude version of the model of mind described in this book) implied that astrocytes *might* play an important part in higher functions, which was then not a generally accepted view since they were usually envisaged to be no more than brain housekeepers. For example, I wrote about the possibility that they have a more important role in 2003 for an online journal (*Science & Consciousness Review*). It wasn't until around five years later that sufficient evidence had accumulated showing that astrocytes *do* play a part in higher functions to allow publication of pioneering papers that attempted to define their likely roles. An especially comprehensive early paper of this sort was published in *Journal of Biological Physics* in 2009 by Alfredo Pereira and his colleague Fabio Furlan from the University of São Paulo. The fact that a preliminary version of the 'landscape' model of mind allowed such an important prediction gives confidence that it may be on the right lines.

The model could have another overall feature worth mentioning; one that may or may not be important. To be honest I don't really know what to make of it, so will simply point it out and leave it at that. Like the prediction of the importance of astrocytes to mind, it is to do with the fact that mind's instantiation in the brain is pictured as dependent on fractals, especially fractal patterns of calcium-ion behaviour. What's potentially interesting, or at least suggestive, from our point of view is that all holograms are fractal. Even though not all fractals are holograms, there is evidently a possibility that, according to our model, some form of holography could be important to mind, and perhaps especially to how it might represent its content.

Speculations that mind and/or its content might depend on holographic principles have been surfacing from time to time more or less since the invention of holography. The distinguished neuropsychologist Karl Pribram (who died in January 2015, aged 95) was especially keen on the idea, which he continued to promote in ever more sophisticated variants until shortly before his death. However, the fact that the idea has been around for so long without getting very far, even though championed by some very able people, suggests either that it is incorrect or that techniques for investigating it are still insufficiently advanced to allow any progress with it. I'd quite like to think, myself, that the idea is good but the techniques are insufficient. After all if nature had the opportunity to make use of holography in brains, which is kind of a neat technique, it would be a little surprising if she didn't make use of it. However, I suppose it could equally be the case that the fractals of our model don't in fact embody holograms, so there's not much point in further speculation about this at present.

Holographic or not, there does seem at first sight to be a degree of complexity overkill in brain organisation. The number of alternative patterns of interconnection that could exist between a brain's component neurons and astrocytes considerably exceeds the number of particles in the visible universe. We have around 90 billion neurons in our brains according to recent estimates and perhaps 100 billion astrocytes. The neurons alone are each said to have an average of around one thousand synaptic connections with other nerve cells and each such connection may have a range of varying 'weights' affecting its strength. Astrocytes, with their 'domains' of two million synapses in humans, can only add to the potential variability. Possible patterns of interconnection are whittled down by anatomical constraints, while synaptic activity is often averaged across multiple synapses both by its inherent stochasticity and no doubt also by astrocyte-enabled averaging. Nevertheless there would seem to be quite enough complexity available in cell interconnectivity alone to provide a sufficient basis for mind. Why then add in another layer of complexity with the fifty or more neurotransmitters and neuromodulators that are known to be relevant to mental function? Surely it would have been far simpler and cleaner from an evolutionary point of view to make do with just one stimulatory chemical and one inhibitor of neural activity. The extra chemicals involved each provides an extra degree of freedom and could, in principle at least, increase functional complexity by a factor of 48! (i.e. exponential 48, which is around 10^{61}). No doubt it's a lot less than this in practice, but there's still a puzzle as to why brains should have evolved in this way.

My dog has provided a plausible answer to the conundrum. She developed a pseudo-pregnancy a couple of years ago when her hormone balance changed and her brain was flooded by a whole new set of neuromodulators. Her mental landscape changed more than did her physical attributes. The basket she usually slept in at night became a den to be protected with fierce growling, especially if any woman came near (she was much less fierce with males, whereas she had previously preferred female humans for company). A little blue toy sheep that she had been given for Christmas became a baby to be cuddled, cosseted and licked. After a couple of hormone injections from the vet to reverse the pseudo-pregnancy she began slowly to revert to normal although it took almost three months for her 'baby' to turn into a scary 'rat' that could be chased and shaken.

Clearly the neuromodulator change had activated a quite specific shortcut in the dog's mind; a change that substituted a landscape appropriate to everyday life for one fit to meet the demands of motherhood. She'd got it wrong that time of course. She's one of those dogs rather prone to getting things wrong albeit with the best of intentions. But normally (i.e. if she'd actually been pregnant) there would have been huge evolutionary advantages to a substitution of this sort. And that's presumably where all the extra chemicals come in. They've been honed through millions of years of evolution to support large changes in mental landscapes; changes that allow instant, or nearly instant, behavioural adaptation to altered physiological or environmental circumstances.

There's another topic I need to mention at this stage: one that is only indirectly connected with landscapes, astrocytes and fractals, though it is to do with a principal means through which brain dynamics connect with social dynamics. It's the so-called 'mirror neuron' system, which was discovered around thirty years ago by Giacomo Rizzolatti and colleagues at the University of Parma. They were studying the activity of single neurons in the motor areas of macaque monkeys' brains when they noticed that some of these neurons became active not only when the monkey they were studying made some particular movement, but also when it saw another monkey making a similar movement. And the effect appeared to encompass more than just information about what the other monkey was doing. It also appeared to depend on the *meaning* of the other monkey's movement since subsequent experiments showed that a mirror neuron connected with reaching for food, for instance, would 'light up' only if the other monkey also reached for food, not if it made a similar movement that was intended for some purpose other than grabbing a banana or whatever.

There has been much debate subsequently about the scope and importance of mirror neuron systems. There's good evidence that we have them also and that they may relate to a range of functions other than purely motor ones. But opinions about their likely role vary from thinking them of marginal importance only, probably mere reflections of the activity of elaborate cognitive systems, to viewing them as a primary and essential source of our capacity for empathy and the like. Clearly, however, they do represent a deep connection of some sort between what's going on in one's own brain and what's going on in the brains of other people. They almost certainly provide the basis for our enjoyment of shows such as *Strictly Come Dancing* and very probably for the national passion for football.

It's especially significant that they reflect meaning. Lots of neurons, in the visual cortex and elsewhere, 'light up' if they spot someone moving in some way, but that's just a matter of information transfer. Neurons that become active in relation to both one's own intentions and those somehow inferred to exist in other people are adding meaning to information. To revert to the metaphor used in Chapter 2, the informational 'rain' has been channelled down a 'valley' in a landscape that represents not only the dynamics of the recipient's brain, but also those of other people. And that channelling is reflected in the activity of mirror neurons. They can be viewed, in other words, as a concrete manifestation of 'extended mind' in action.

This chapter has been all about looking at the machinery behind mind, which turns out to be quite unlike any automaton from the Enlightenment or android from the twentieth-century imagination. Charles Sherrington's 'enchanted loom' analogy was perhaps the closest approach hitherto to what it is like, but even that isn't very close. Mind is in fact dependent on a maelstrom of chemical and electrical activity in the brain shaping itself into the evanescent forms that we have envisaged as 'landscapes', though cloudscapes in a storm might have been a more apt analogy in some respects. The shaping depends on memories, some derived from genetics, with origins extending back to the dawn of life, others formed on the basis of regularities that occur in our physical environments; the majority, perhaps, originating from our social environments – the cultures we live in, our nations, workplaces, families and friends. Our minds are systems that resonate with their immediate environments but are formed by the history of life and humanity as well as our own personal histories.

The huge downside of our type of mentality lies in its tendency to get stuck in configurations that lead to unrealistic and ugly outcomes, as witness the sad history of fanaticism down the ages and its modern embodiments in all sorts of different guises ranging from the Taliban to Scientism. Sleep, so effective in restoring landscape flexibility to individuals, is relatively powerless against fanaticism because the relevant 'landscapes' extend into social, group dynamics where the most effective counter agents to excessive fixity are not sleep but death and democracy. These have their downsides too, for death removes people who might help to restore good sense and vitality along with its elimination of individual fanatics, while democracy fosters sudden mass surges into unwise enthusiasms and witch hunts along with its encouragement of variety, adaptability and realism. Fortunately it's not all gloom and doom; the tendency of social landscapes to rigidify also enables the maintenance of useful institutions – schools, universities, orchestras, the list is endless.

None of this, however, tells us where consciousness comes from. The 'mind' that we have pictured could all be occurring in the dark, so to speak, with no appreciation of the red of a rose, the beauty of a sunset or the light in a lover's eyes. Our picture so far cannot even account for the painfulness of pain, let alone the awareness of being a self. What makes waking life so essentially different from our mode of existence during dreamless sleep entirely escapes all that has been said up to now. We'll therefore be leaving landscapes, fractals, calcium ions and the rest of the 'machinery' behind in succeeding chapters, regarding them as no more than entities subserving the *content* of consciousness. They are not directly responsible for the *existence* of consciousness, so far as we know, though clearly the two do interrelate in *some* way. Finding potential solutions to these 'existence' and 'interrelationship' problems is crucial to reaching any full understanding of human nature, so we'll be focusing on them in what follows in the hope of at least being able to delineate the outlines of reasonable answers and of finding evidence to indicate which particular answers may eventually turn out to be the most reasonable. Before doing that, however, there's a centrally important feature of both our objective and subjective worlds that we need to tackle.

4
ON TIME

I need to digress a bit at this stage and talk about time, for reasons that should become clear in Chapter 6. In any case time, and especially its passage, is of central importance to the world and all our experience of it. Without time there could be no dynamic processes and hence no minds of any sort, so we need to try to gather as many clear ideas about its nature as possible. People often launch into this topic by paraphrasing St Augustine's remark that he knows what time is until asked to explain it! We can actually do a bit better than him these days because we can say for sure what time *isn't* and also say something about how it appears to behave, though only in relation to particular contexts.

What time *isn't* is the way we instinctively think of it. We automatically picture it as some sort of universal process like a river's steady flow – 'time, like an ever-rolling stream, bears all its sons away' – which carries us along from a remembered past towards an unknown future and sooner or later tips us into the grave. It's 'the old enemy' indeed. The present moment converts an unreal future into a reality that quickly fades into a semi-real memory.

Isaac Newton formalised this intuition with his picture of the present moment as an interface existing everywhere between future and past that steadily progresses from the past into the future at a rate of one hour every hour (which doesn't make much sense, if you think about it, because it's like trying to define a speed through space as one mile every mile). According to his mathematics, as expressed in what we now call differential calculus and he called 'fluxions', the present hardly exists because it is of infinitesimal duration. Bishop Berkeley, an early critic of this idea, called infinitesimals 'the ghosts of departed quantities', which was quite apt in its way even though they have since proved enormously useful in maths and science.

Newton's picture embodied ideas about the present moment and about the flow or 'arrow' of time that need to be tackled separately. We also need to keep a clear distinction in mind between time as conceived by physicists and time as experienced by us. Let's try to deal with the physical concept of a present moment first. We know for just about certain that Newton was wrong about its universality. Albert Einstein's relativity theory has been tested to all sorts of limits and has never yet failed. It shows beyond reasonable doubt that what one observer may see as simultaneous events in his/her 'present' can appear to another observer as separated in time. Moreover the other observer can see one of the events (say event A) as occurring *either* before *or* after the other event (B), depending on how he/she is moving relative to the events. The only caveats are that the second observer must *be* moving relative to the events and must be further away from at least one of the events than the events are from each other.

What this means is that the 'present moment' has to be regarded as a purely local concept, not a universal happening. In physics it is often best regarded as referring to the moment of some causal interaction, though this isn't always satisfactory because it implies that 'now' can have hugely different durations. The 'now' of a photon hitting my eye, for instance, would be tiny (less than a femtosecond), whereas the 'now' relating to support of a paving slab by the ground it is lying on could be years. It's probably best in physics to forget ideas about 'the present' or 'now' and think instead of measures of time, causal interactions and their direction (i.e. why causes precede outcomes).

The universe can be thought of as a vast assemblage of clocks measuring time since every particle within it has a frequency, and frequencies (once a day, for example, or twenty beats per minute or 10 megahertz) are inherently dependent on the passage of time. However, relativity theory also shows us that measures of time (seconds, hours or whatever) are interchangeable with measures of distance (inches, metres, light years etc.) because the speed of light is always constant; it's the same in all reference frames. Quantum theory carries a similar message, for the frequency of particles is equivalent to their wavelength. Which particular measure or combination of measures, whether of duration or distance, is most useful to a physicist varies with contexts, especially ones to do with relative states of motion. So the fact that all those clocks are ticking away everywhere doesn't mean that they are measuring some constant, universal flow of time. They are just measuring their own little local spatio-temporal circumstances.

Nevertheless time is directional in a sense, at least as far as we and the classical world in general are normally concerned. Particles encounter one another, interact in some way, and an outcome emerges. They appear out of the future, interact in their own shared 'present' and leave a 'memory' of their interaction to propagate forward in time. Why isn't it the other way round? Why do causes precede their effects, in other words, and not effects precede their causes? The equations of physics are almost all fully reversible in time (albeit with complications, which I won't go into here, when it comes to something called 'parity'), so why don't they work backwards as well as forwards? There are answers to this question but it also turns out that they too don't apply universally. In some special circumstances time can indeed appear to go backwards.

Time's arrow is often attributed to the fact that, in the universe as a whole, disorder (entropy) almost always increases. This is because disordered states are always far more likely to develop than ordered ones unless there are complex mechanisms in place, such as are found in living organisms, for producing order. Even then it's only possible to generate order by causing greater disorder somewhere else. After all it is vastly more probable that a gust of wind will scatter the seeds away from a dandelion clock than that it will sweep up scattered seeds and fix them to the dandelion. The fact that the plant managed to grow an ordered seed head in the first place wasn't down to gusts of wind but to work that it put into producing order, achieved at the expense of converting sunlight into more disordered forms of energy. There's a similar reason given by quantum theory for the direction of time's arrow. Causal interactions are equivalent, at a fundamental level, to the 'quantum measurements' that were mentioned in Chapter 1. You can get a definite outcome from the potentialities in some pre-measurement state, but it is absolutely impossible to get from that outcome back to the pre-measurement state. You may be able to get *another* pre-measurement state on which further measurements can be made later on, but there's no way of getting back to the original.

Hence causes must always precede outcomes as far as 'measured' causes and 'measured' outcomes are concerned (you may recall from Chapter 1 that 'measurements' are always reciprocal in that what gets measured is also measuring its measurer, which sounds complicated but simply means that the measuring process is circular). However, there's a loophole here. As far as pre-measurement states are concerned, there's no reason for supposing that time has any preferred direction or even that there *is* any time. And indeed the 'delayed choice' experiments mentioned earlier show that apparently (i.e. from our perspective) *earlier* behaviour of a hitherto unmeasured quantum particle can be affected by a *later* (again from our perspective) measurement made on it.

Some physicists have supposed that this could be down to 'backward in time' effects. They may have been influenced by Richard Feynman's suggestion that positrons (positively charged particles) can be thought of as time-reversed electrons (which are negatively charged) and by a proposal (i.e. the so-called 'transactional interpretation' of quantum theory made by John Cramer) that some of the mysteries of quantum theory become less mysterious if one supposes that 'waves' can be of two types, either 'advanced' or 'retarded', which means that some may travel backward in time and others forward. It's perhaps simpler, though, to suppose that time is irrelevant to the pre-measurement world, rather as space appears irrelevant when it comes to pre-measurement 'entanglement' between particles (see Chapter 1).

There is a difficulty with supposing that time could be non-existent or take some very different form in the pre-measurement world, but it's a problem that may be more apparent than real in a rather literal sense. It is to do with the fact that the pre-measurement world is usually represented in terms of wave functions, which, being waves, evolve in clock time. The so-called 'quantum Zeno effect' appears to confirm this picture. The Zeno effect is a term for the experimentally confirmed fact that one can prevent or at least greatly delay the decay of radioactive atoms, for instance, by observing them (i.e. by measuring their state) at very short intervals, in the megahertz range. This is usually put down to there being insufficient time between successive observations for the wave function of the atom to evolve from 'no decay' to 'decay', so the atom gets stuck in a non-decayed state. However, since measurements are reciprocal, it could equally be the case that the Zeno effect is down to there being insufficient time for the *measurer* to evolve into an 'I see decay' state. In other words, it could be a passenger in a train situation. The usual interpretation of the effect is to suppose that the measurer is stationary and the wave function is flowing past its observational 'window', but maybe what's actually going on is that the measurer is the mover and the wave stationary. We, in our role as measurers, could be moving past a stationary timescape. This is perhaps the most straightforward way to conceive of 'delayed choice' findings, which appear to be down to change in the measurer's temporal perspective on a holistic, atemporal pre-measurement situation.

And that's basically what contemporary physics has to say for reasonably sure about time. There are all sorts of further, wildly divergent speculations about what might be going on in the foundations of reality, zones that we know nothing about. Julian Barbour, for instance, suggests that if we could but see 'reality' for what it truly is then we wouldn't find any time at all but only a vast assemblage of objects called 'relative state spaces'. Lee Smolin has proposed, in contrast, that time is 'real' in a way that space isn't. Barbour has pictured a static universe of unimaginably baroque complexity. Smolin has envisaged a reality of constant change that spans temporal successions of universes. Barbour is Parmenides to Smolin's Heraclitus. St Augustine, familiar as he was with the views of both philosophers, would have found nothing too unfamiliar in the modern physicists' contrasting views of time's essential nature – or its illusory nature according to Barbour.

But the more recent developments, the things that we could honestly relay to St Augustine as being probably true and that might be news to him, are that 'now' doesn't mean very much in physics except as a reference to the occurrence of some particular causal interaction, while time's arrow (i.e. the directionality of causal interactions from cause to effect and thus from past to future) is down to the irreversibility of 'measurement' processes on microscopic scales and the universal tendency to increasing disorder on human or larger (i.e. 'thermodynamic') scales. Incidentally, though disorder is often thought to be 'bad' because of its consequences for our lives, we do need it. Highly ordered states are like ice: rigid and sterile. Add some disorder to ice and you get water with its innumerable potentialities for variety and life. Of course a bit more disorder still produces steam, which isn't so good, though it does have its uses. We do need time's arrow and its ever-escalating disorder, despite our grave-ward destination!

Time as we experience it, however, is nothing much like physics time except for its directionality, pointing from past through present towards the future. Even that differs from physics because we seem to move at wildly varying speeds through time; a day in childhood can seem to last as long as a week in later life, while an hour when you are busy may pass like a few minutes when you are bored. Physics time, however, chunters along pretty steadily unless you're looking at relativistic speeds, and even then time would seem to pass steadily from the point of view, if it had one, of the relativistic object (unless it was a photon that is; from a photon's 'point of view' time wouldn't appear to pass at all!). The really big difference between physics time and experienced time, however, is that the present moment means something definite to us – it is what we exist in, in so far as we *are* our experience of existing. Our brains are constantly imagining, and making predictions about, the future but we don't *experience* it until it has arrived. We may remember the past, but the experience of remembering is generally quite different from that of experiencing the present. At least that's the normal state of affairs although the 'flashbacks' that traumatised people sometimes get can seem as real, or even more real, than the original experience. Also, as we'll be seeing later, the distinction between present and future isn't always quite as sharp as is generally supposed.

The present moments that we experience (William James, the great American philosopher/psychologist, called them the 'specious present') can vary a lot in clock-time duration from around one-tenth of a second if you are very aroused in some way, to several seconds if you're relaxed and listening to soothing music, for example. They can perhaps be regarded as arising from a summation of huge numbers of individual, physical, 'causal presents' in the brain. This would account for why we share the same directionality of time with physics, while the varying durations of present moments and varying speeds through time that we experience could be put down to details of the summation process and how we remember it. If there is this sort of summation, which seems pretty certain, something must bind all the 'physical' moments together into a *single* experiential moment. What could bind them together temporally is one of the big mysteries of neuroscience, along with the mystery of how it is that spatially separated events in the brain get integrated into what we experience as single percepts, single feelings and so forth. We'll take a look at possible answers to the mystery in succeeding chapters. In the meantime, I'd like to describe a couple of oddities about our relationship to time.

The first oddity is highlighted by an illusion called the moving coloured dot illusion. If people are shown a dot of light that is very briefly flashed at one position on a screen and is followed one-fifth of a second later by another dot flashed at another position, they experience seeing only a single dot that apparently moves from the first to the second position. Then comes the really strange bit. If the first actual dot is coloured red and the second actual one green, the moving dot that people see appears to change colour from red to green about half way across the screen – i.e. about one-tenth of a second *before* the green dot has truly appeared. People can see something before it happens! Or so it would seem. And 'seem' is the operative word here.

What the illusion really shows is that all those ephemeral 'physical' causative moments out there are summed up over a duration of around one-fifth of a second *before* they appear in an 'experiential' moment. People's brains do a whole lot of work on the momentary physical happenings before a person gets to experience them consciously. The colour illusion results from averaging out the colour contributions of the two actual dots to the perceived one and then putting the perceived change of colour in what would have been the appropriate position if the two stationary dots had really been a single moving dot. What the illusion shows is that brains can somehow change the apparent timing of 'physical' causative moments before producing an experiential moment. But they don't need to foresee the future in order to do this because experiential moments always lag a fraction of a second behind the 'physical' moments that contribute to them.

The second oddity is more than just an illusion, although there are plenty of people wishing that it were nothing more who try to ignore or debunk it. It goes by various names – 'pre-sponse', 'presentiment', 'feeling the future' – and, though distinctly undramatic in itself, is totally at odds with the current world view of 'man a machine' aficionados. Here's what it boils down to: in psychological experiments, testees are often given some stimulus or other procedure that produces a measurable response. For example, if people are shown an unpleasant picture, one of their responses takes the form of an increased tendency to sweat, which shows up in a change in the electrical resistance of their skin (called the Galvanic Skin Response – GSR). They don't give the same response if they are shown a picture of a relaxing scene instead. It's a measure that has been used for sixty or more years in lie detector tests, as well as in laboratory tests. What's remarkable is that the same pattern of responding can be seen, albeit in very muted form, up to around three seconds *before* the stimulus is given. There's no way this can be explained on anything like the same basis as the coloured dot illusion. It looks like a genuine 'backward in time' effect.

Of course many dispute that it *is* genuine, often quite angrily. The (alleged) effect seems to have been first mooted, almost in the form of an online urban legend, around twenty years ago. Well-designed tests of it have been carried out since. Perhaps the best known of these, because it was especially thorough and got published in a prestigious journal, was one conducted by a senior academic psychologist from Cornell University named Daryl Bem. He arranged to conduct nine different types of psychological experiment that have been in widespread, almost routine, use and are known to produce measurable effects of one sort or another. He carried them out in the normal manner, using randomisation, control stimuli and the like, with the exception that he looked for the anticipated effects *before* giving the stimuli. And he found them – although in one of the nine types of experiment the 'pre-sponse' seen didn't reach statistical significance, which it did in the other eight.

Some ninety experiments to test for the reality of the effect have been reported, at the time of writing, from thirty-three different laboratories spread among fourteen countries. Combined, using a technique called 'meta-analysis', they give overwhelming evidence that the effect is indeed real. Critics often suggest that experiments giving negative results (i.e. ones that *don't* find the supposed effect) are much less likely to get reported, so evidence from meta-analyses is likely to be biased in favour of positivity and may thus be spurious. However, statistical techniques (funnel plots) are available nowadays to check on this possibility. It seems likely that only nine unreported negative findings exist in relation to pre-sponse experiments (some of the *reported* findings were negative of course, like the one out of the nine in Daryl Bem's study), while over *five hundred* unreported negative experimental outcomes, in addition to those already reported, would be needed to negate the positive findings. That's far more than could plausibly exist. The pre-sponse effect, weird as it is, has been established beyond reasonable doubt even if, sadly, not beyond unreasonable doubt.

It's worth emphasising that these 'pre-sponses' are small compared to the normal responses you get in the relevant experiments, and quite often can't be seen at all. There's a bit of statistics called an 'effect size', giving an estimate of the strength of any effect. Whatever produces the 'pre-sponse' has an effect size of around 0.2. It means, roughly speaking, that the unknown 'something' that produces the 'pre-sponse' contributes only around 4 per cent of the total of causative influences resulting in the measured outcomes (i.e. only 4 per cent of the 'variance'). Although 4 per cent both sounds and is a small proportion among all the other mechanisms that generate a GSR, for example, which would include autonomic nerve impulses, sweat gland responsiveness and so on and so forth, lots of well-accepted, 'normal' causal influences in psychology and medicine have similar effect sizes (e.g. different educational methods producing different levels of examination success or the effect of particular prophylactic medications on the likelihood of getting some disease). Question the reality of the pre-sponse effect and you have to question the validity of large swathes of the sciences of psychology and epidemiology among others, which is certainly an option but not one that many of Daryl Bem's critics would welcome!

There *are* plenty of critics, by the way. This is one of the few contexts in which unjustified biases can be found in the most apparently respectable of websites. Some of them are expressions of the sort of helpful scepticism that contributes so much, and is essential to the progress of science. Most, however, seem to be down to people who apparently feel personally threatened by evidence that scientific understanding of the world is incomplete, and who resort to propaganda and misinformation in defence of their world view. They are a bit like the late nineteenth-century writers, so impressed by the development of thermodynamics that they over-interpreted it, and wrote things such as 'the earth cannot be more than twenty-five million years old' or 'heavier than air flying machines are impossible'. One can well imagine that, had the internet been around in those days, a quick Google search would have turned up lots of pages alleging that the Wright brothers had faked their flight in order to advertise their bicycle business! Many of the anti-Bem comments have a similar flavour.

Unfortunately we've no idea what the pre-sponse does imply, other than that events in the very near future can affect what happens in the present to a small extent – which is what upsets the critics since it runs counter to mainstream scientific beliefs of the last two hundred years or more. There are some indications that the effect is stronger when the test of it involves fast, unthinking or reflex activity. Presumably, in tests that are more 'automatic', there's less room for additional causal influences on measured outcomes, so the retro-causal effect size can be proportionately larger. There's thus no evidence that the effect has any connection with consciousness of what is going on at the time of the pre-sponse and some evidence that it doesn't. So far as I know, there's equally no firm evidence that pre-sponses depend on later conscious awareness of stimuli or other procedures that retro-cause them, although unconscious awareness of the stimuli does appear to be a prerequisite (any attempt to bypass this restriction would have to be complicated at best).

Does this pre-sponse oddity imply an inherent 'fuzziness' in the nature of time, or an uncertainty in the direction of its arrow that is somehow amplified by brains? Is it something that is a property of minds only and, if so, does consciousness play a part in producing it? These questions remain entirely open, telling us that we are still not very much better off, after all, than was St Augustine when it comes to explaining time.

5
NEW FRONTIERS

Towards the end of the twentieth century there was a resurgence of interest in the nature of consciousness, such as had not been seen since mid- to late Victorian times. But it took a very different form from its predecessor for it grew from roots in analytic philosophy, neuroscience and quantum theory, sources far divorced from the mesmerism and spiritualism that had inspired the earlier movement. There were two principal ideas behind it: the first being that consciousness is an emergent property of neural activity; the second, less popular and considerably vaguer, was that it might have something to do with quantum theory.

Many proponents of the first approach (the so-called 'eliminative materialists') at first supposed that consciousness would ultimately prove fully reducible to neural activity and we should then be able to say something like 'my nerve cells are in state XYZ' that would be wholly equivalent to some other statement of mine such as claiming that 'you look lovely today'. They had carried the 'older scientific orthodoxy' picture described in Chapter 1 to its logical conclusion. But it soon became clear that this approach, if taken to such an extreme, was a busted flush for a whole lot of compelling philosophical reasons, many of them articulated by a young Australian philosopher named David Chalmers in the mid-1990s. His notion of what he dubbed the 'hard problem', the question of how 'qualia' (i.e. the qualities that we consciously experience such as pain, redness, the scent of a flower etc.) could possibly arise from brute matter, has remained influential ever since.

Meanwhile neuroscientists were producing ever more evidence that both the content and even the apparent occurrence of consciousness does vary along with neural activity of particular types or in particular parts of the brain. The philosophers therefore came up with a fudge somewhat reminiscent of Descartes' pineal gland proposal. They suggested that neural activity might have two separate properties: consciousness and also what the neuroscientists could see with their EEGs, fMRIs and direct recordings from nerve cells. Which property manifests (i.e. whether a conscious experience or alternatively a recording of an electrical field, for example) was regarded as dependent on whether the neural activity is viewed from the inside or the outside – i.e. from a 'subjective' or an 'objective' point of view. This was dubbed 'property dualism'. The question of exactly what distinguishes 'subjectivity' from 'objectivity', given that *all* experience is ultimately subjective in the sense of belonging to some individual person, was usually swept under the carpet. Others preferred the very similar idea that it is *information*, not neural activity per se, that can have two aspects depending on whether it is seen from a first-person or a third-person point of view. Instead of a pineal gland to solve the conceptual problems, you have a 'property' or an 'aspect'. Clearly this is an explanation that explains nothing, much like the pre-Newtonian habit of ascribing the weight of objects to their 'ponderousness'.

Not surprisingly, panpsychism – the view that *everything* might be 'conscious' in some sense or other, which seems to have been around since the dawn of time – began to regain popularity in the present century, albeit hesitantly and tentatively. It faces the obvious problem of how to conceive of the 'consciousness' that would have to be attributed to a rock and the less obvious one of how a lot of little panpsychist experiences in nerve cells or bits of them could link up to give our sort of complex but unified experience. The latter was a particularly acute form of the 'binding problem' that also faces all neuroscientific theories of consciousness; namely, how is it that separate events occurring in different parts of the brain or at different times can blend into a single conscious episode?

However, problems to do with both what to make of the experience of rocks and how binding could occur can be shelved, for a time at least, if one goes for some form of what Bertrand Russell and others have termed 'neutral monism' – namely the idea that the basic stuff of the universe, whatever that may be (and according to Immanuel Kant and many others, we can never actually know its true nature), consists of matter and conscious mind in equal measure. It has a good deal in common with some theological conceptions of God the Father as 'the ground of all being'.

The most radical type of neutral monism was advocated in the late 1940s by Wolfgang Pauli, the sparky and spiky genius who originated the 'Pauli exclusion principle' of quantum theory, and Carl Jung, the well-known psychiatrist and guru. They formulated what has since been termed (by Harald Atmanspacher and his colleague Wolfgang Fach) the 'Pauli/Jung conjecture'; the proposal being that the basis of reality is an unknown something in which consciousness and matter form an indivisible unity. Bertrand Russell's version of neutral monism, in contrast, on the other hand, seems to have assumed that the two entities retain some of the qualities that we attribute to them while confined within his concept an all-encompassing monism. The Pauli/Jung proposal is probably still the best that we can achieve in this particular connection, so it is the concept that I shall run with in some of what follows. Although it might appear at first sight to be no great advance, or even no advance at all, on property dualism or dual aspect theory, it has potential implications that can take us a good deal further, as I shall hope to show you in Chapter 6.

But I'm getting way ahead of myself here, for there is a major impediment to any discussion of the nature of consciousness that we need to address; namely that we can only discuss experience that we can recall having experienced. If we can't remember it, we can't discuss it. Similarly, scientists can only investigate *reportable* consciousness; that is they can only investigate reports, verbal or behavioural, made by humans of their experience, or behaviours reflecting what experimenters infer to have been conscious experience in the case of animals.

It's been generally assumed since the days of Sigmund Freud that a lot of mental activity is unconscious; a view that seems to have been confirmed subsequently by psychological experiments involving 'masking', 'unconscious priming', 'blindsight' and the like. However, we also know that there isn't a sharp cut-off between conscious and unconscious mind. The 'it's on the tip of my tongue' phenomenon that we've all experienced shows that one can be consciously aware of pre-conscious mental activity. Similarly so-called 'fringe consciousness', for example, the vague awareness that one is starting to feel cold while concentrating on chatting to a friend, blurs the boundary. Thus, when we talk about 'consciousness', maybe we're just talking about some small part of it that happens to have got into our memories.

Luckily, although there's always liable to be doubt about where unconscious mind ends and conscious awareness begins, and even more doubt as to how best to conceive of the nature of any foundational, panpsychist 'consciousness', the situation isn't as bad as it may appear when it comes to our own form of experience thanks mainly to the pioneering work of Benjamin Libet, a twentieth-century neurophysiologist. He used a range of techniques to convincingly show, in work that has been subsequently confirmed by others, that information reaching the brain and destined for consciousness (which most of it isn't; estimates of the proportion of incoming information that enters consciousness generally put it at <10 per cent) takes about one-third of a second to gel into conscious experience. Since it has to be held in mind for that length of time, his finding means that consciousness and early stages of memory processes must be intimately interrelated. It looks, therefore, as though consciousness *is* a memory-related phenomenon. Not all of it moves on into permanent storage, but nevertheless it seems that we don't after all need to worry that our form of conscious experience might be something disconnected from memories of it. The fact that we can study only *reportable* consciousness is thus a less worrying limitation than might otherwise have been the case.

Some philosophers have actually built a theory of consciousness on considerations of this sort, the so-called Higher Order Thought (HOT) theory. It comes in several varieties, but the basic idea is that consciousness *is* an expression of the interaction of an unconscious experience with the 'thought' (which may itself also be unconscious) that 'I am having this experience'. Like many similar arguments, however, it carries a whiff of tautology and probably tells us more about the semantics of consciousness-related words than about the phenomenon itself.

Arguments about hypothetical thought experiments, often centring on semantic attributions, are almost always inconclusive. A particular favourite of philosophers has been one involving 'Grey Mary', an imaginary, all-knowing neuroscientist brought up in a world entirely without colour. Would she experience something new – something that she had been entirely unable to imagine or predict – when first shown a red rose? This question, originally asked by Frank Jackson in 1982, has never yet received a definitive treatment. Jackson's own first answer was 'yes, she would experience something she couldn't have predicted' on first seeing the rose, but he has subsequently changed his mind. Papers are still frequently written to 'show' that the answer is probably 'yes' – or probably 'no' as the case may be. The *Journal of Consciousness Studies* alone gets around six erudite papers on the 'Grey Mary' question sent to it most years now, and there were more than that ten years ago (only a small proportion pass peer review and get published). So let's move on to see what empiricism, not semantics, has to say about consciousness. We'll take neuroscience first.

On the whole, neuroscience is driven by available observations and techniques. The last time there was the sort of interplay between theory and experiment that is often seen in physics was over a hundred years ago when Ramón y Cajal was having his dispute with Italian neuroanatomists about neuron connectivity. He won the argument with his detailed anatomical studies although, as so often happens, it now looks as though both parties were right since gap junctions, mentioned in Chapter 3, provide the sort of syncytial structure that the Italians had envisaged, existing alongside Cajal's more easily detectable synapses. Because the field is technique driven one needs to keep in mind that it is especially prone to over-interpreting positive findings. Experiments designed to provide Popperian refutation of some theoretical prediction are quite rare and theory is usually playing catch-up with experiment, which is far from ideal.

There are two related sources of difficulty particularly liable to lead to over-interpretation or even misinterpretation of observations. One, already noted by the nineteenth-century neurologist Hughlings Jackson, is down to the fact that most neuronal activity is inhibitory – i.e. it damps down the activity of other neurons. Therefore, if some part of the brain is damaged or otherwise interfered with, it can be difficult to know whether observed outcomes are due to loss of a positive function mediated by the affected area or loss of an inhibitory effect on other areas and their consequent overactivity (or vice versa, of course, if some local area is stimulated rather than impaired).

The other interpretative difficulty that can cause problems generally is down to the fact that the brain is a 'small world' network. Any two randomly selected neurons are likely to be connected to one another by only a few intermediates; probably fewer than the famous 'six degrees of separation' said to characterise the social networks that allow connection of any two randomly selected people anywhere in the world. David DeMott, an EEG researcher, remarked rather sourly as long ago as 1970 that 'one can find connections between any two randomly selected points in the central nervous system . . . given enough time and a goodly research grant'. Yet both the prevalence of inhibition and the degree of interconnectivity are sometimes ignored by experimenters when it comes to explaining the alleged significance of their findings.

Nevertheless there is plenty of good evidence that particular brain areas do specialise in particular mental functions. Most of the basic evidence came from neurologists studying stroke victims or people with brain injuries; the First World War was especially 'useful' in this respect as not all bullet wounds to the brain were fatal and it was often fairly obvious which bit of some unfortunate soldier's brain had been injured (albeit not usually the extent of injury). The end of the twentieth century brought tomographic brain scans, especially fMRI and, to a lesser extent, PET, which allowed discovery of functional localisations of the same type that the neurologists had picked up but with far greater sensitivity and discrimination. Scans can be used in uninjured people, which is a huge advantage of course, and provide detailed 3D images of which bits of the brain appear to be doing what.

It's worth recalling what fMRI studies involve and what they measure. They involve surrounding people's heads with very powerful electromagnets so they can hardly move, to the accompaniment of loud bangs of almost jackhammer intensity. Perhaps surprisingly, people can follow instructions to undertake mental tasks in these circumstances and the apparatus measures small increases (of the order of 10 per cent), coinciding with these tasks, in a measure (blood oxygen saturation) of the metabolic activity in localised brain areas. These increases are averaged over periods of about a minute and can be localised down to volumes of as little as a cubic millimetre (though it's worth also remembering that each cubic millimetre may contain around 10^5 nerve cells and as many astroglia, so the localisation is still fairly coarse by the standards of the brain). The increases probably reflect extra energy needed to pump more ions across cell membranes, which implies that fMRI does actually measure 'mental' activity according to the view of 'mind' offered in Chapter 3, not just some remote correlate of it.

The details of localisation are mainly of interest to specialists, and their implications are in any case far from clear in the context of a 'landscape' picture of mind since zones of *reduced or unchanged* activity may contribute as much to a landscape as zones of increase. It's the overall shape that matters. The situation is like the one that applies in binary logic where a '0' (no increase in energy usage) may convey as much or more information than a '1' (increase in energy usage) depending on the relevant expectation values. However, two findings are of more general interest. The first of these is the negative one that neither consciousness nor mental function in general appears to be affected by the very large static magnetic fields involved in fMRIs. In contrast, much smaller but changing magnetic fields, as used in Trans-cranial Magnetic Stimulation (TMS), do affect the content of mind and consciousness via stimulation or inhibition of neural activity (which effect is seen depends on the frequency of the applied magnetic field). TMS probably acts via direct effects on transmission of impulses along axon bundles. The lack of effect of static MRI fields provides another confirmation that mentality is all about dynamics, and the same probably applies to consciousness.

The other overall finding is to do with the malleability of localisations. At one time, thanks to over-interpretation of findings on 'split brain' patients, the left hemisphere came to be variously regarded (at least in the popular literature) as the conscious 'boss' of the brain, or alternatively as a sort of stand-in for Mr Spock, rigid, logical and uncreative, while the right hemisphere was cast either as a creative genius unfairly suppressed by the left or as a semiconscious handmaid to the left. And it is generally true that, in right-handed people and most left-handed ones too, language functions are mainly on the left so it's easier to elicit verbal reports of experience from that hemisphere, especially if it is disconnected from its partner on the right as happened in the 'split brain' cases. Moreover the left specialises to some extent in sequential tasks other than language and in happy emotions, while the right tends to deal in spatial gestalts and less happy emotions.

However, fMRI studies suggest that none of this is unalterable. For instance, professional translators employed by the United Nations have been shown to use their *right* hemispheres for their native language, keeping the left for additional languages. Likewise professional musicians tend to use their left hemispheres for their performances, while amateurs mainly use the right. It seems highly likely that the shifts occurred as a result of training because similar if less dramatic ones have been shown to result from practice or accompany recovery from injury. There are no longer any good grounds for supposing the left hemisphere to be inherently more 'conscious' in some sense than the right, though the two may, or occasionally may not, make differing proportional contributions to aspects of the *content* of consciousness.

EEGs, which have been around for eighty years, have provided another important investigatory tool. They measure brain electrical fields and thus ion shifts in real time (up to around 200 Hertz) but with a spatial localisation accuracy of only cubic *cent*imetres. They thus provide a picture of temporal dynamics that is around 10^4 times as discriminating as fMRI, but are about 10^3 less discriminating spatially. They are complemented to some extent by the more recent technique of magnetoencephalography (MEG), which mainly provides a measure of current flows along bundles of axons.

The most important insights they have provided are that mentality in general, along with both the content and the presence of consciousness, are reflected to some extent in the power of particular EEG frequencies and their distributions across the brain. Deep sleep, for example, is accompanied by high-amplitude slow waves spreading across the brain from front to back; dreams by faster, choppy activity in cerebral hemispheres; strong emotions by increased 'theta' (3–7 Hertz) activity; cognitive activity by increased power in 'beta' (14–25 Hertz) frequencies. It also turns out that different areas of the brain that are engaged in dealing with information about some percept tend to show synchronised EEG activity in the 'gamma' (25–80 Hertz) band. For example, if you see a brown bird flying onto your bird table, gamma activity in your colour- and motion-sensing areas (which are separate) will tend to synchronise, soon followed by activity in your 'that's a bird' identifying area, which is several centimetres distant from the primary visual ones.

When first discovered, it was claimed that gamma synchrony associated with perceptual integrations of this sort is always at about 60 Hertz, but subsequent work has shown a much wider frequency distribution. Many researchers like to think that the synchrony solves the 'binding problem' (i.e. how it is that activities in different brain areas combine to form single percepts). 'Neurons that fire together are together' is the mantra often offered. However, the jury is still out on this because a number of apparently well-designed, well-conducted experiments have been reported that *haven't* found the effect. The 'landscape' picture of mind actually allows a more nuanced view. Particular 'valleys' in the landscape may often manifest in synchronised ion shifts, but they don't necessarily *have* to do so. Therefore it wouldn't be surprising if negative findings sometimes occur and there's no need to blame them on incompetence of the experimenters responsible for them – a manoeuvre that, regrettably, has been seen from time to time.

Another popular focus of research has been on recordings from single neurons, mainly in animals, especially macaque monkeys and rats, but only occasionally in humans for obvious reasons. Many neurons can be found that become active in relation to a single type of stimulus. David Hubel and Torsten Wiesel got the Nobel prize in 1981 for the first such discovery, which was on cats not rats. They found cells in the visual cortex that responded best when the animals were shown lines tilted at some particular angle. Subsequently all sorts of other specific responders were found, some, referred to as 'Grandmother cells', that would light up on seeing a specific face, such as one's grandmother's; it was even claimed by one group that they had found a 'Bill Clinton neuron' at the time of his presidency. Quite a cottage industry has grown up over recent years to deal with hippocampal neurons. The hippocampus is an area of the brain that has a range of memory-related functions and also deals with orientation and spatial location, so that neurons can be found that are active only when an animal reaches some particular part of a maze it has learned, for example.

Perhaps the main general lessons to be learned from these neuroscience techniques is that specialisation of function in the brain occurs on all sorts of scales, both spatial and temporal. Specific neural correlates of the content of both mind and consciousness can be found over scales spanning at least six spatial and nearly two temporal orders of magnitude. However, there appears to be no special determinant of the *existence* of consciousness, as opposed to its content, other than that brain activity of *some* sort appears to be a prerequisite. Changing tack at this stage to look at drug effects rather than measures of neuron activity promises to tell us a bit more. The two most relevant types of drug are general anaesthetics and psychedelics. We'll take them in that order; both pose mysteries but potentially enlightening ones – or at least they ought to prove enlightening.

The main impediment to gaining enlightenment from looking at general anaesthetics is that there are so many of them with few characteristics in common apart from lipid solubility. One particularly good anaesthetic, the noble gas xenon, is actually chemically inert, suggesting that anaesthesia is unlikely to be down to chemical binding so much as to some mechanical effect such as being the wrong shape in the wrong place, so to speak. The current fashion is to suppose that they are effective by virtue of blocking stimulatory synaptic activity, enhancing inhibitory activity or blocking voltage gated ion channels. Indeed a range of different anaesthetics have been shown to possess effects like these in differing measure, some being apparently mainly inhibitors of glutamate (stimulatory) neurotransmission and others mainly enhancers of GABA (inhibitory) activity.

There are evidently 'chicken and egg' questions here – do the changes in synaptic activity cause, result from or merely correlate with loss of consciousness? As these questions are largely unanswered at present, one can't draw any firm conclusions about how the drugs work, let alone infer anything definite about the basis of consciousness. However, it's probably fair to say that brain energy requirements appear always to be reduced under anaesthesia, indicating that there's less ion pumping going on. This, in turn, suggests that a certain amount of neural activity is needed for the *existence* of consciousness, or at least for the ability to remember it and thus experience human-type awareness. Unfortunately, though, general anaesthetics don't (yet) tell us anything definite about the *sort(s)* of neural activity that's needed.

Psychedelics pose a different range of mysteries as they don't abolish consciousness, indeed they sometimes appear to enhance it, but produce profound alterations in its content. Unfortunately research on them has only recently and stutteringly resumed after an almost fifty-year hiatus due to the 'war' on drugs. This particular war was at least as counter-productive as nearly all others, the main harms being the huge enrichment of actual criminals and the criminalisation of very large numbers of relative innocents. The promising research on psychedelics that had got under way in the late 1950s was a minor, but nevertheless significant, casualty of friendly fire; significant particularly because it might, if it had continued uninterrupted, have weakened the many misconceptions about the nature of mind that cropped up in the late twentieth century due to computer analogies.

If I were to tip some tea over the laptop I'm using to write this, it would probably stop working; it certainly wouldn't suddenly start displaying T. S. Eliot's poem *East Coker*, for example, on its screen. Yet giving someone a psychedelic drug can do something very like this. The drug gets into their brain, interferes with its circuitry and they suddenly start having fantastical but highly organised experiences that sometimes replace experience of what is actually going on in the world around them or sometimes become integrated with reality. Benny Shanon, an Israeli psychologist, has given us an especially detailed account of the extremely varied experiences that can be produced by the hallucinogenic brew ayahuasca. The vividness and range of the experiences that he identified are remarkable, though some particular themes were relatively common such as seeing hallucinatory people, snakes, animals or elaborate 'historical' scenes.

Most psychedelics affect mainly 5-hydroxytryptamine (serotonin) neuromodulation in complex ways that are poorly understood. Others, however, such as ketamine (which is also a general anaesthetic in high dosage) or salvia divinorum, don't appear to have any primary effect on serotonin or its receptors but do affect a range of other receptors, especially the NMDA subtype of glutamate receptor in the case of ketamine, which may account for its anaesthetic properties. There seems to be quite a lot of overlap between varieties of experience produced by typical and atypical psychedelics, as shown by comparisons that have been made, for example, between effects of DMT (dimethyltryptamine, a serotonin mimic and a principal ingredient of ayahuasca) and ketamine. The experiential complexities are so great as to defy detailed analysis but can be understood in general terms as due to profound modifications of mental 'landscapes' that may be outcomes of a wide range of causal processes.

These modified 'landscapes' can be very odd indeed. To take examples that Benny Shanon recounted, one of his informants described having become fully immersed in scenes from a 'celestial palace' when he was looking upwards, but could also clearly see the hall that he was in when he looked downwards. That sort of dissociation has parallels perhaps in several areas of mental life. Other experiences are more puzzling. For instance, Europeans have reported seeing 'jaguars' after ingesting their brew during sessions in their home countries; South American Indians who have never left their tribal areas have viewed 'Russian cities' with their onion church domes; a labourer with only three years of primary education saw 'mathematical formulae which he understood as conveying the basic laws of the universe'.

The question raised by experiences of this sort is 'where do they come from?'. We have thought of mental 'landscapes' hitherto as based on neural memories with a range of genetic, personal and socio-cultural origins. One might expect, when normal function is disrupted by a psychedelic drug, that the more deeply engrained memories would come to the fore to mould hallucinations and other experiences – hence perhaps the frequency with which snakes are hallucinated since fear of them probably has a genetic basis.

But Europeans are unlikely to have had much experience of jaguars – surely they ought rather to hallucinate leopards, lions, tigers, cheetahs or whatever; sometimes these creatures do indeed appear, though less often than black pumas. Jaguars are about equally likely to be seen as the African big cats with which Europeans are far more familiar. There's evidently a puzzle as to where the jaguars could have come from. Similarly Amerindians are unlikely to have had any profound experience of anything remotely resembling a Russian city, or an uneducated person any great involvement with mathematical equations. It looks, therefore, as though our 'landscapes' may have determinants more complex than has been envisaged hitherto. There is a problem here that's worth noting, but noting it is about all that can be done at present; it conceals a hidden essence of creativity, perhaps, that's likely to remain hidden for a good many years yet.

The overall message about consciousness given by empiricism is that 'mentality' is a function of a somewhat compartmentalised brain, aspects of which enter conscious awareness. *How* they do so has mainly been left as an open question, one that involves attention (and thus short-term memory) but whether in a primary or secondary role remains unclear. There's ongoing, but unresolved and complicated, philosophical and neuro-psychological debate over whether attention is necessary for consciousness, whether unconscious attention can occur and whether unattentive ('un-intentional' in philosophical parlance) consciousness exists. My personal feeling is that the issues involved are largely semantic and won't be sorted out until new concepts and definitions are introduced into the debates. The question of *what* aspects of mentality become conscious, however, has been given a number of relatively clear and convergent answers.

An especially influential answer to the 'what' question was formulated in the 1980s by psychologist Bernard Baars with his description of a 'global workspace'. The idea is that lots of little specialist brain modules are continually both doing their own thing unconsciously and competing for access to the workspace. When a module gains access, its information is distributed to all the others and, in the process, acquires consciousness. The aspect of mentality that is conscious at any given time is thus the content of the global workspace. It's a nice picture in some ways but has a big problem in that access to the workspace is envisaged as being controlled by attention – how could attention possibly 'know' where to allocate access *before* information in a chosen module is available globally?

This question was never answered convincingly. It was often swept under the carpet in discussions of the theory that tended to centre on the fact that it says nothing about *why* workspace content should become conscious. Nevertheless the concept of a global workspace is still considered useful in some rather general sense, but it is usually regarded nowadays more as an emergent function than as the sort of thing-like arena that Baars envisaged. Psychologists at the time he was writing were enamoured of clunky charts with many boxes, the more the better, labelled with tags such as 'executive function', 'attention', 'perception', 'global workspace' and so on. Modelled on engineering flow charts, the boxes had arrows pointing between them, sometimes bidirectional, sometimes solid or dotted. Despite being almost always quite misleading in one way or another, flow diagrams of this sort can, unfortunately, still be found in some textbooks.

A rather similar idea, but one that approaches the issues from a different angle, so to speak, has gained popularity over recent years: namely Giulio Tononi's 'integrated information' concept. This grew from Gerald Edelman's suggestion that consciousness is a function of 're-entrant' neural activity where information in separate brain areas is constantly updated via reciprocal interactions with one another. Edelman (who died in 2014) was a Nobel prize-winning physiologist who collaborated with Tononi in later life. Tononi now suggests that consciousness is something that emerges quantitatively as informational integration between neurons increases. He has expressed this notional integration mathematically, which adds to the appeal of the idea in neuroscientific circles. Instead of a 'global workspace' conceived as an arena from which information is broadcast brain-wide, Tononi offers a picture of extensive, changeable neural alliances co-ordinating their informational content. It fits well, intuitively at least, with the empirical observation mentioned earlier that separate brain areas engaged in processing information about particular percepts often develop synchronous (gamma-band) activity.

The 'landscape' picture also encompasses a global workspace-like notion since the largest landscape features present at any particular moment, the dominant attractors, channel incoming information (what we envisaged earlier as 'raindrops' falling on the landscape) and relate it to information inherent in the landscape form (i.e. to the memories from which the landscape is constructed). The 'workspace' at any particular time *is* the dominant landscape feature. There is also an idea of 'informational integration' implicit in this picture since incoming information is both 'integrated' by virtue of being channelled down some particular 'valley' and is related to the stored information that determines the structure of the 'valley'. The latter step is the one that automatically adds meaning to any information-theoretic concept of incoming information. It's a step that has to be added later, in a less natural way, in Tononi's model.

Other advantages of the 'landscape' picture include its prediction of the important functional role of astrocytes and its accounts of both the necessity for sleep and the apparent naturalness of tool use, all of which are outside the remit of the 'integrated information' picture taken in isolation. Both 'landscape' and 'integrated information' pictures lack, however, any intrinsic account of the *why* of consciousness; they merely suggest that *what* becomes conscious is, respectively, activity related to the currently largest 'landscape' feature or that showing maximal informational integration. To get to the *why* they must appeal to philosophical concepts of property dualism or dual aspect theory, which, as mentioned earlier, are far from satisfactory since they are basically fudges adopted for the sake of 'saving the appearances' in much the same way and for the same reason that epicycles were added to the diagrams of Ptolemaic astronomy.

Given the reluctance or inability of mainstream neuroscience to engage with the essence, as opposed to the content, of consciousness, many have wondered whether quantum theory might offer a way forward. As noted in Chapter 1, physicists in the mid-twentieth century were already wondering whether consciousness might *determine* the manifestation of reality. By the 1980s a good many people were developing ideas about how brain and quantum might interact. John Eccles (a Nobel prize-winning neurophysiologist) and Karl Popper (the great philosopher of science), for instance, collaborated on a theory that entailed a concept of the existence of quantum theoretical influences on the probability of release of packets of neurotransmitter molecules from pre-synaptic terminals in the brain. By the 1990s there was a plethora of such theories, most of which didn't follow Eccles and Popper's idea but focused instead on the possible occurrence of hypothetical, long-lasting quantum coherent states in the brain. There were so many theories of this type that several volumes would be needed to do full justice to them all. I'll briefly describe only the one that has proved to be the best developed and most long-lived as a representative of all. It's the so-called Orchestrated Objective Reduction (OrchOR) theory due to Stuart Hameroff, who is an anaesthetist, and Sir Roger Penrose, the well-known mathematician and physicist.

First, a few remarks on the 'quantum coherence' at the basis of the theory. This is a 'pre-measurement' (see Chapter 1) state that was originally conceptualised in terms of Bose–Einstein (BE) condensation. A BE condensate is an assemblage of bosons (i.e. particles or quasi-particles with integer 'spin' such as photons or the electron pairs of low-temperature superconductivity) that have been induced somehow, often by reducing their temperature to near absolute zero, to share a common lowest 'ground state' and thus a *single* wave function. In effect such condensates become a single giant particle. Nowadays, with ever-increasing understanding of the universality of pre-measurement 'entanglement', the concept of coherence has been applied more widely to all systems that have become entangled and remain so, however widely separated their parts may appear to be situated from our point of view.

The main problem is that, unless the systems are held at temperatures close to absolute zero or are composed only of particles such as photons that don't interact with one another, they decohere extremely fast (usually on time scales of the order of 10^{-15} seconds or less at room temperature) due to 'measurement' by their environments. People wanting to create quantum computers are currently putting enormous efforts into maintaining coherent states of one sort or another between as many particles as possible for as long as possible. So far as I know, the current record is something like coherence between sixty particles maintained for a millisecond, but the pace of research is such that the record is probably being broken as I write.

Now back to OrchOR theory: it has two main pillars. The first concerns microtubules and their properties. These are molecular structures within all cells including neurons that have a wide range of functions that can sometimes be brain-like albeit on very small scales. Stuart Hameroff has pointed out that the single-cell organism *paramecium* can show quite complex behaviours despite lacking a nervous system, which may well depend on its microtubules. The tubulin molecules that compose microtubules can adopt one of two different shapes, each having different functional consequences. Quantum coherence involving effects on tubulin shape might in principle allow some form of quantum computation that could be scaled up via microtubular functions to affect neuronal function.

The second pillar has to do with Penrose's idea, described in detail in two books and a number of papers, that 'collapse of the wave function' (a concept that overlaps to some extent with the more recent notion of decoherence due to measurement) depends on a gravitational criterion and that collapse is associated in some way with consciousness conceived as a 'non-computational' phenomenon. Combining the two pillars allowed calculation of the number of tubulin molecules likely to be involved in any particular episode of coherence (around 80,000 according to the most recent estimate of which I am aware), located consciousness in an aspect of quantum theory (namely the 'collapse of the wave function' process) and gave an account of how it could influence neurons. It achieved a lot, in other words, within a neat, if complex, package.

Sadly, although OrchOR theory has never been definitively refuted and remains consistent with a range of neuroscientific findings, there are doubts about its viability, doubts that apply with as much or more force to all (so far as I know) the 'quantum consciousness' theories that were proposed at around the same time. The main problem is that they were formulated before the idea of decoherence due to environmental 'measurement' became firmly established in the mid-1990s. This allowed calculation of the rate at which a process, equivalent for nearly all practical purposes to the notion of 'collapse of the wave function', occurs. Previously, apart from Penrose's suggestion about a gravitational criterion for collapse that has never gained wide acceptance, people had no idea of how to calculate rates for the decoherence equivalent (i.e. 'collapse of the wave function') that they used. Estimates had ranged from instantaneous to many milliseconds. As it turns out, decoherence happens over time scales many orders of magnitude too brief for pre-measurement coherent states to be considered as possibly offering a plausible basis for consciousness, or at least not for our sort of conscious experience.

Remaining advocates of these theories have been encouraged by recent evidence that photosynthesis and enzyme activity depend on quantum coherence, along with electron 'tunnelling', which is also a quantum phenomenon. However, these phenomena occur over times of around 10^{-12} seconds and intra-molecular spatial distances only, scales that are about ten billion times too small to translate into any sort of plausible basis for consciousness in the brain. Despite their apparent promise and allure when they were first proposed, it now looks as though the 'quantum consciousness' theories of the 1990s may have been heading off down a blind alley. Indeed it is quite possible that theories of the earlier, Eccles and Popper type could have been have been a bit nearer the mark when it came to looking at how quantum theory might relate to conscious experience.

The overall picture that we've arrived at so far shows consciousness to be dependent on the more extensive and elaborate aspects of brain representations of 'mind'. There's quite detailed information, thanks to brain-imaging techniques, about which parts of the brain contribute most actively to particular aspects of the content of consciousness. Early stages of memory processes are closely involved with the sort of consciousness that we experience, while memory in general, interacting with incoming information from the body and the world at large, in a very real sense *is* 'mind'. We are creatures created from memories of all sorts (biological, personal and socio-cultural) that resonate with the ongoing dynamics of our bodies and environments.

When it comes to accounting for the existence of consciousness, however, we're still very much in the dark. Neuroscience, on the one hand, can explain it only by appealing to a philosophical fudge (i.e. dual aspect theory or the like). Attempts to base it in quantum theory, on the other hand, have so far failed to live up to their initial promise. In the next two chapters I want to outline an example of a type of theory that is better able to account for the fact of conscious experience and that may have a chance of telling us something about its true nature.

6
BROKEN SYMMETRY

> [Conscious] mind ... is simply the intrinsic temporality of a physical event.
> *(Alfred North Whitehead)*

This chapter will explore possible implications of a big assumption, which is that the reality underlying all appearances is a unity of the type envisaged by Wolfgang Pauli and Carl Jung. They termed it the *unus mundus*, the ground of existence. It can be conceived as pure potentiality, lacking any of the attributes or appearances that manifest in our world. To that extent it is perhaps like a pre-measurement state in quantum theory but is more profound than any wave function, deeper even than the quantum theoretical notion of a 'wave function of the universe', for the potentialities in pre-measurement states relate to so-called quantum 'observables' only (things such as position, momentum, energy, spin). Those in the *unus mundus*, in contrast, relate to everything including Galilean secondary or 'subjective' qualities, and no doubt also to aspects that wholly elude our mundane experience and observation. But when it comes to positively describing the *unus mundus*, one can only echo the Christian mystic Meister Eckhart, or a Buddhist sage, and say: 'It is not this, not that.'

However, there is one thing that can be said for sure about the *unus mundus* concept; something that endows the idea with *genuine* explanatory value and thus distinguishes it from notions of property dualism or the like.

What's certain is that symmetries must be broken when the unity is broken to allow manifestation of potentialities within it. From our point of view the most basic and important symmetry that's broken is the one between 'subjective' and 'objective' worlds. We live existences in which these two aspects are entirely distinct. Conscious experience is always subjective, never objective – which is basically why life has been so difficult for neuroscientists for they can study only objective *correlates* of reported consciousness.

Symmetry is chief, perhaps, among the few notions that are built into the foundations of contemporary physics. It crops up in all sorts of contexts. For instance, there's a theorem (Noether's theorem) proving for sure that important conservation laws are dependent on translational symmetries. Conservation of momentum and angular momentum are expressions of the indifference of physics to moving to a different position and changing spatial orientation respectively. In other words, the underlying physics will be the same whether it occurs in Cape Town or Tokyo, facing north or facing west, and those 'symmetries' entail the two varieties of momentum conservation. Of course, if the surrounding circumstances are different in different places, if the ambient magnetic field is not the same in Cape Town as in Tokyo for instance, then physical outcomes may differ; nevertheless the underlying physical processes don't change. Similarly, conservation of energy is related to the indifference of physics to clock-time temporal transitions. Nature will behave in the same way at tea time as it does at lunch time.

Other conservation laws involve more esoteric symmetries. For example, conservation of electric charge depends on something called 'gauge invariance' – roughly the indifference of physics to the metric used to make physical measurements. Indeed the very existence of electromagnetism, according to the gauge theories that underpin present-day understanding of forces and particles, is down to nature's 'need' to maintain a quantum property called $U(1)$ phase symmetry. Incidentally it's worth pointing out that these gauge theories are more than flights of fantasy since they have been used to predict the existence of previously unknown particles that have subsequently turned up in high-energy experiments. The theories really do tell us something about the true nature of nature, and symmetry is all-important to them.

If symmetry is important to physics, so too is broken symmetry. As mentioned in Chapter 1, for instance, the split in an original electro-weak unity that occurred in the first few moments of the universe was responsible for almost all the variety of the objective world that we live in. But broken symmetries also have much more mundane relevance. They are broken, to take just one instance, when steam condenses leading to manifestation of the more restricted symmetries of water, which, in turn, may break in favour of the even more restricted symmetries of ice crystals. Perhaps the main lesson to be learned from this is that ice is a potential inherent in steam, one that may or may not manifest depending on the ambient temperature of the environment. Even the electro-weak split is thought to have occurred when the energy density or 'temperature' of the infant universe had dropped sufficiently to allow the symmetry to break.

There are thus two main questions to ask about our hypothetical subjective/objective symmetry split: firstly, what is the ambient circumstance that enables splitting and secondly, precisely where and when does the split occur? Regrettably, neither I nor anyone else, so far as I know, has any good ideas to offer about the circumstances that might maintain unity or enable splitting. Temperature is almost certainly irrelevant here but it's hard to imagine what might take its place. But there is quite a lot that can be said in answer to the 'where and when' question. We need to gather ideas about the most likely location of the split and symmetry considerations themselves offer a big, hard-to-miss hint as to the best place to start looking.

The hint depends on the curious absence of a particular symmetry, albeit an absence that seems to have attracted very little attention hitherto. There is a form of symmetry in quantum theory between spatial position and momentum, which are 'non-commuting' variables (i.e. share a Heisenberg uncertainty relationship) and are also both quantum 'observables' (i.e. are aspects of the 'objective' world). Now energy and time are also non-commuting variables and energy is a quantum 'observable' like momentum but time isn't; a particular position in time is not like a particular spatial position in this respect and there is thus a mismatch with the otherwise apparently comparable momentum/position symmetry.

Surely this is telling us that, while energy is part of the 'objective' world, there is a sense in which time, unlike spatial position, isn't part of that world, in which case it is likely to be an aspect of the 'subjective' side of the split. However, that's not to say that time isn't 'real'. The Heisenberg uncertainty relationship with energy implies that it must be as real as energy itself, which is probably the most 'real' entity that we know of, for the 'virtual particles' that play such essential roles in quantum field theory and have been experimentally proven to exist (via the Casimir effect) owe their existence to their very brief temporal manifestation that allows their energy to vary over a wide range. The flipside of this energy–time relationship has important implications for 'subjectivity' that we'll come to soon.

It looks, in brief, as though energy and objectivity are on one side of a divide, while time and subjectivity are on the other. This picture fits, in a general sort of way, with the distinction between 'physics time' and 'experiential time' that was discussed in Chapter 4. The split suggests that 'physics time' or 'clock time' is a sort of notional metric derived from experience of subjective time, which is the primary phenomenon. This is a complete reversal of the Newtonian concept of the basis of time, where it is pictured as entirely 'objective', an aspect of the framework needed, indeed, for the very existence of objectivity. This assumed, essential 'objectivity' carried over into relativity theory with only minor modification but the theory does actually incorporate a hint that the assumption could have been incorrect. Time has to be given the opposite sign to space in relativity theory; that is if space is positive, time is negative or vice versa (it doesn't matter which way round the negativity is assigned). This surely provides a strong indication that time is utterly different from space in *some* way, despite our ingrained habit (dating back to Hermann Minkowski's famous 1908 lecture) of lumping the two together as 'spacetime'.

An implication of all this is that any subjectivity/objectivity split is likely to coincide with energy 'measurements'; each measured packet of energy that manifests in the 'objective' world will be accompanied by a tiny packet of 'subjectivity'. Such packets have been dubbed 'qualions' (by Tal Hendel and others). Eccles and Popper used the term 'psychon' to refer to their rather different concept of elementary packages of 'consciousness', but perhaps that's another word that could be hijacked and applied to the notion offered here. A problem with both of these terms, however, is that they may suggest something particle-like, along the lines of a photon, electron, neutron or whatever, but they actually refer to something totally unlike any 'objective' particle. I, therefore, prefer to call them 'scintillae of subjectivity' (SoSs). Individual SoSs won't be anything like our own familiar conscious experience if only because they must lack the memories that imbue our experience with meaning. Each can be envisaged as bearing much the same relationship to our experience as does a single calcium ion in the brain to the physical basis of our minds.

One of the important properties of SoSs is that it is possible to attribute an 'objective', clock-time duration to each of them because of the Heisenberg time/energy uncertainty relationship. If the 'objective' energy is very precisely measured (by its environment) then the associated SoS will be of relatively long clock-time duration. For instance, if the uncertainty in the 'measurement' of an energetic event is only 10^{-33} joules, the associated SoS will have a clock-time duration of 0.1 seconds. This energy is actually tiny (less than the kinetic energy of a smallish molecule at a temperature close to absolute zero), but it's the *uncertainty of measurement*, not the actual total energy manifesting, that matters.

It's highly probable that sufficiently precise energy manifestations occur to allow lengthy SoSs, just as sufficiently short clock-time durations are possible to allow manifestation of virtual particles. However, macroscopic events in general will have large energy uncertainties; so large that any associated SoSs will be of almost infinitesimal clock-time duration. An implication of this is that one need not worry about any 'subjectivity' attributable to rocks or even to raindrops falling – nor indeed to nerve impulses firing since these also have large energy uncertainty – just as there is no practical need to bother about the virtual particles that are constantly popping in and out of existence, suffusing all of space around and within us.

Our brains, however, are packed with a huge variety of energetic events happening in ordered patterns. Many of these are likely to manifest very small measurement uncertainties; for example, ion bindings to some specific protein or events involving phonons (i.e. quasi-particles of vibrational energy). SoSs with clock-time durations of as much as 0.1 seconds are thus likely to be ubiquitous in brains.

From its own point of view any individual SoS simply exists in a present moment. It has no way of measuring the clock-time duration of that moment. A duration can be assigned to it only from an external, objective viewpoint. Nevertheless SoSs in large numbers can be envisaged as mapping the patterns of energetic happening with which they are associated. In a loose sort of way, they can be said to convert spatio-temporal events in the brain into a tempero-spatial format. Objective maps in the brain, those that manifest as unconscious 'mind', are founded on the reality of space and its changes over time; subjective ones can be thought of as founded on the reality of time and its changes over space according to this picture.

The picture also suggests that parts of our bodies other than our brains have to be viewed as harbouring their own primitive forms of 'subjectivity' too. However, since reportability of any such forms of consciousness, if this is ever possible, is always going to depend on brain activity, they must be indistinguishable from brain consciousness for nearly all practical purposes. Occasional reports by highly trained athletes and sports persons of their 'body consciousness' may suggest that making a distinction isn't necessarily absolutely impossible, but it's never going to be an easy call to make, let alone to investigate.

The flow of conscious experience can thus be viewed as much like the flow of a river. Molecules of water get together in a river by virtue of forces between them and are channelled through some geological landscape. Similarly, SoSs associate because of their clock-time temporal overlap and are channelled through a mental 'landscape' of the type envisaged in earlier chapters. Viewed in this light, the temporal and spatial 'binding problem' (that of how it is that unified experience can arise from events apparently separated in time and space) that has so exercised neuropsychologists is seen to be actually a pseudo-problem.

★★★

In a way this chapter has taken us full circle back to Descartes, which is hardly surprising given that it follows his strategy of splitting the world into two. But it's Descartes with a difference. *Res extensa* is now an 'objective' world that is basically a manifestation of energy in its many forms, forms that provide energy with spatial extension and allow application of a temporal metric. It includes the unconscious aspects of mind that are embodied in the material world, principally in our brains. *Res cogitans* has become a 'subjective' and conscious condition founded in temporality, one that is able to map aspects of unconscious *extensa* mind into a format that provides subjectivity with temporal extension and allows application of a spatial metric, thanks to the energy/time relations existing in brains.

The picture can be summarised as:

(*Objectivity*/**Space**-*time* :: *Subjectivity*/**Time**-*space*)

I need to be a bit provocative at this stage — some might say a lot more than 'a bit' — for reasons that we'll get to soon. I need to point out that this picture has at least as much in common with the 'spiritualist' concept described in Chapter 1 as with the new (scientific) orthodoxy one. Here they are again for comparison:

New orthodox:

(*Extensa* :: *Cogitans*)

Spiritualist:

(***Extensa****/cogitans* :: *Extensa/**Cogitans***)

The 'new orthodox' picture posits a simple divide between *extensa* and *cogitans*, whereas the 'spiritualist' one implicitly mixes them up to some extent, and so does the view advocated in this chapter. The advantage of the new view is that it broadly indicates the *whats*, and even to some extent the *hows*, of mixing. This, in turn, allows a search for actual evidence that might be relevant to the notions described and to their refinement. However, the fact that it has commonalities with the 'spiritualist' outlook suggests that one may need to look for evidence outside the 'orthodox' box. To make a start in this direction, one could probably do a lot worse than retrace some of the steps taken by late nineteenth-century enquirers into consciousness. I'll be trying to make this sort of start in Chapter 8.

Why not look first in a more orthodox (by contemporary standards) direction? The main reason is that purely 'objective' investigations are unlikely to offer unambiguous evidence of the sort that's needed. They can look at only one side of the coin, so to speak. For instance, as I was writing this chapter, a report appeared (in the *Journal of Cerebral Blood Flow and Metabolism*) by a group who had been trying to differentiate patients with minimally conscious states from others with vegetative (i.e. completely unconscious) states, both caused by brain damage. Rates of glucose metabolism (i.e. energy generation) were down on average 42 per cent in the minimally conscious cases and 55 per cent in the vegetative ones, mainly in fronto-parietal areas of the brain. Interestingly enough, metabolic rates in brainstem and thalamus, although down from normal, were no different in the two groups, which is a finding hard to reconcile with some older, neuro-emergentist theories of consciousness that attributed it to feedback loops between cortex and brain stem.

'Nice!', I thought on first reading the report. Surely this is a good indication that energy generation and consciousness are directly linked as SoS theory requires. Then I quickly realised that one can readily come up with perhaps twenty reasons to account for the association that have nothing to do with SoS theory. I had initially succumbed to one of the besetting sins of contemporary neuroscience: that of over-interpreting evidence to fit one's theoretical prejudices. It's an urge hard to resist. Better perhaps to look instead for types of evidence that are hard to accept! And the nineteenth-century investigators certainly had a knack for doing exactly that, as we shall see in Chapters 8 and 9.

In the meantime it's worth identifying principal questions about which evidence is needed. There are perhaps two particularly important ones, though a whole range of others could be asked, plus a very big hole in the theory that needs to be plugged if possible. Let's take the questions first.

The first has to do with the fact that it is not at all clear what it means to say that time is principally a 'subjective', consciousness-related phenomenon. What does this imply for the status of consciousness and in particular for memory (with which, as we saw earlier, our form of consciousness appears intimately related if only from an 'objective', *extensa* point of view)?

The second question has to do with a Newtonian principle that has never been in much doubt, namely that actions are generally associated with reactions. We've pictured how *extensa* might mould *cogitans* via the generation and channelling of SoSs in the brain. How does any back action of *cogitans* on *extensa* manifest itself? One can certainly predict from the theory that back action of *some* sort can be expected because 'subjectivity' has been viewed as being just as real as 'objectivity' despite occupying a partially separate realm of existence. But how any back action might manifest is an entirely open question. Answering it holds out the prospect of telling us a lot about the nature of consciousness.

Evidence relevant to these questions would boost confidence that the theory is on the right lines and might help towards its refinement. Lack of evidence would tend to refute it. Before going on to look at what evidence is available, however, we need to tackle the enormous hole in the theory, an omission to do with the fact that it says nothing about what is to us the essence of conscious experience – namely that it is built of qualities: the redness of red, the painfulness of toothache, the beauty of a rose etc., to give clichéd examples. SoSs can't individually provide these qualities (*qualia* in technical parlance), if only because they lack the meaning that is intrinsic to qualia. What individual SoSs are envisaged as doing is to provide the 'subjectivity' that endows qualia with consciousness, in much the same way that we envisaged calcium ions as providing a 'physical' basis for unconscious mind. No individual calcium ion could supply the meaning essential to mind; similarly no individual SoS could supply the quality of a quale.

How then could qualia arise? I should confess straight away that I can't offer any firm answer to this question and thus plug the hole in the theory in a wholly satisfactory way. What can be done is to delineate the hole a bit more precisely and offer ideas about how it *might* be filled. That's the topic of the next chapter.

7
QUALIA

This chapter centres on yet another split: not the one due to René Descartes this time but an earlier distinction made by Galileo Galilei, which has been foundational in science ever since. He described it thus:

> Now I say that whenever I conceive any material or corporeal substance, I immediately feel the need to think of it as bounded, and as having this or that shape; as being large or small in relation to other things, and in some specific place at any given time; as being in motion or at rest; as touching or not touching some other body; and as being one in number, or few, or many. From these conditions I cannot separate such a substance by any stretch of my imagination. But that it must be white or red, bitter or sweet, noisy or silent, and of sweet or foul odor, my mind does not feel compelled to bring in as a necessary accompaniment. Without the senses as our guides, reason or imagination unaided would probably never arrive at qualities like these. Hence I think that tastes, odors, colors, and so on are no more than mere names so far as the object in which we place them is concerned, and that they reside only in consciousness.[1]

---

1 It turns out that 'consciousness' was a mistranslation of *corpo sensitivo* (sensible body) according to Filip Buyse in a talk given at a philosophy seminar in Quebec in 2013. This rather suggests that Galileo himself may not have viewed secondary qualities as being quite so 'imaginary' as has usually been assumed subsequently.

Hence if the living creature were removed, all these qualities would be wiped away and annihilated. But since we have imposed upon them special names, distinct from those of the other and real qualities mentioned previously, we wish to believe that they really exist as actually different from those.
(Translation by Stillman Drake, Discoveries and Opinions of Galileo.
New York: Anchor Books, 1988)

It's a quotation that has come to occupy an almost *Magna Carta*-like place in the story of science. Galileo's odours, colours and so forth have subsequently come to be termed 'secondary qualities', while his 'real' qualities – shapes, motions and contiguity are the examples he gave above – are called 'primary qualities'. This terminology has been useful for it has encouraged study of *quantifiable* metrics but it is deeply misleading in relation to *qualities*. What more accurate distinction(s) should Galileo have made?

Motion, for instance, is both a primary and a secondary quality. One can both cognitively grasp a concept of motion and directly perceive it; there is 'something that it is like' (see Thomas Nagel's concept of consciousness as present when there 'something that it is like' to be that something) to perceive movement by an object and this is an aspect of perception that can occasionally be lost following a stroke or other brain injury. We don't perceive weight so directly. Galileo's distinction holds a bit better in relation to that for there is only 'something that it is like' to perceive feelings of strain in muscles or pressures on the soles of our feet. To get to a 'quality' of weight in the abstract we have to imagine something measured by our bathroom scales or go to an abstract idea of inertial mass as in the relativistic formula $m = e/c^2$. But there is still 'something that it is like' to have these cognitions. Archimedes' famous 'Eureka!' moment provided an especially dramatic example of what it can be like to have a cognition.

It looks, therefore, as though Galileo's distinction should have referred to qualities directly perceived by our perceptual systems versus those inferred by our cognitive systems but he didn't intend anything of that sort – or at least he has subsequently come to be interpreted as having said something different (see footnote). What he is thought to have meant by a 'real' quality wasn't a quality at all; it was an inferred aspect of the world to which a quantifiable metric could be applied.

Can we suppose then that Galileo's distinction should have referred to one between a perceived quality of any sort, and a measurable, inferred aspect of nature? Well, this might have worked in the sixteenth century but not nowadays. Perceived qualities themselves are evidently aspects of nature and, since the development of perceptual neuropsychology, it has become normal to measure them or at least to measure the simpler aspects of reports made of some of them, for example, colour saturation or sound intensity. Decibels are as much units of a metric as are kilograms. Where Galileo was correct was in attributing qualities to brains; they belong with 'subjectivity', not with 'objectivity'. Given the argument offered in the last chapter, one may wonder if he was as correct in his assumption about their ephemerality but that's a question that I'll postpone dealing with for later.

He also said that 'odors, colors, and so on are no more than mere names' apparently because 'they reside only in consciousness', thus possibly implying that qualities themselves are nothing more than cognitive constructs. This was an idea that was taken up by twentieth-century psychologists and given its strongest expression in the Sapir–Whorf hypothesis: the theory that language determines what we think and perceive so that speakers of different languages might have differing perceptions. This idea was tested quite thoroughly, the principal conclusion being that the hypothesis is misleading and that qualia are described, but not determined, by language. There is actually some evidence that infants can make perceptual discriminations, presumably on the basis of qualia that they experience, which they may later fail to make if their native language does not support that type of discrimination. The loss is probably analogous to loss of ability to pronounce phonemes that don't figure in their native language, which occurs as they grow up. It can be taken as a further indication that qualia are constructed in some sense by the brain, which may lose the ability to form specific qualia if they are never practised. That's no indication that qualia are not 'real' however. They are features of a realm of reality (i.e. the subjective realm) that is to an extent distinct from the 'objective' world but they are as real within their own realm as was the stone within his objective world that was kicked by Dr Johnson when he wanted to refute philosophical idealism.

Given that qualia are real, it would be fair to ask: what are they? The short answer is that they are states of consciousness that, according to the thesis advocated in this book, are assemblages of SoSs formed by a mental 'landscape'. This merely re-states the theory already offered and a potentially more fruitful question is to ask what differentiates one quale from another. Why isn't red experienced as blue, for example, or as the sound of a bell or even as a feeling of unease? We know that overlap between qualia can and does sometimes occur. Particular colours do tend to go along with particular moods or sometimes more specific feelings. Most of us have probably encountered some ghastly shade of paint that makes us feel nauseous, for instance. Then there are the cases of synaesthesia that have been thoroughly studied by Vilayanur Ramachandran in California, for example, or Semir Zeki in London. They come in many varieties that can affect all perceptual modalities so far as is known. The commonest is to perceive particular letters or numerals as being coloured in some particular shade despite their being printed in black on white as far as the rest of us (and the light absorption characteristic of the 'objective' pigment used) are concerned. These people still experience the meaning of 'A' but may invariably see 'A' as yellow, say, rather than black. Experiencing some particular sound in relation to seeing a particular shape is among the many other types that have been reported. And careful investigations have shown that the reports of synaesthetic experience are veridical, not down to 'imagination'. These hybrid qualia, too, are 'real' within their own realm.

In a crude sort of way, one can say that people with unusual brain wiring can experience unusual qualia or connections between qualia. But this still leaves the question open of what normally differentiates qualia. It merely adds further questions to do with how and why they can sometimes overlap, though it does reinforce the other evidence that they are generated, or at least formed somehow, by the brain. After all, there are no obvious differences between the nerve impulses, the ionic waves or whatever in one part of the brain that could differentiate such phenomena from those in another part of the brain. Distinctions between qualia must derive from something quite complex and subtle.

It could be claimed that qualia derive their distinctiveness somehow from characteristics of mental 'maps'; from the brain's representations of its environment and internal states – from its dynamic 'landscapes' – in other words. And this is no doubt true but doesn't take us much further since it simply shifts the problem. It tells us nothing about how such maps could acquire the necessary distinctiveness or what form the maps could take. They're certainly not lines on a sheet of paper, but they must own some mapping characteristic equivalent to lines on a sheet of paper. What brain characteristic could correspond to a line on paper? One might suppose that it would be as easy to draw a line in a waterfall as one in the constant fluxes occurring in a brain.

Nevertheless *something* allows very clear distinctions between an experience of pain, say, and that of happiness, or one of seeing a colour in contrast to hearing a gong. Some of the neurotransmitters associated with experience of happiness and pain are different, as are the anatomical areas of brain involved in differing experiences. Yet there's also a lot of overlap too. It would be hard indeed to account for the stark differentiation between red and blue solely in terms of neurochemistry or anatomy, though these factors are no doubt contributory.

Giulio Tononi, proposer of the 'integrated information' view of consciousness described earlier, has suggested the notion of a 'qualia space', analogous to the notional 'colour space' that describes the different hues we are able to experience. Qualia are defined by their 'position' in relation to the different dimensions of the space. This provides a different and far more sophisticated approach to the problem of qualia differentiation than thinking of it in terms of questions about what could draw lines on a map. It raises questions instead about what might contribute dimensions to the space. In the case of 'colour space' there are three dimensions down to the three types of colour receptor in our retinae. In the case of 'qualia space' one can well suppose that anatomical variations contribute several dimensions, neurochemical variations others, with yet more dimensions contributed by factors harder to identify.

It's an attractive idea. The main problem with it, so far as I can see, is that it's hard to envisage how one could get from entities in a 'qualia space' of that sort to the conscious qualia that we experience. On a pure 'integrated information', dual aspect view of the origin of consciousness there is no obvious reason why informational integration in one part of such a space should lead to an experiential (as distinct from behavioural) outcome any different from integration in another. Adding in our 'landscape' picture of what may underlie integration helps a bit since it attaches meaning to information and meaning is intrinsic to qualia. Given the evolutionary origins of many 'landscape' features, the picture can also be used to account for some features of some qualia: the sheer nastiness of pain, for example, and perhaps the joy of sex. Psychologist Nicholas Humphrey has even provided a nice evolutionary story about why red is red (in a book titled *Seeing Red*).

All the same, Tononi's 'qualia space' is a representation of an essentially 'objective' entity, deriving its dimensions from 'observables'. How could it, or at least its content, translate into 'subjective' format? If one attributes translation to supposed 'dual aspects' of information within the space the problem of why information in one part of the space should differ experientially from that in another would seem to remain. If, alternatively, one supposes that the whole space translates somehow into a subjective format there would seem to be a problem to do with how an entity conceived spatially could survive translation into an apparently non-spatial format. The SoS concept, or something similar, might come to the rescue here since SoS fields are conceived as having a spatial aspect. Their spatiality is, however, viewed as a property secondary to their primary temporality and it is therefore not clear that they could support translation of an 'objective' qualia space.

Despite the attractions of the qualia space notion, I suspect that some deeper bridging principle is needed to account for qualia differentiation; something with a foot in both objective and subjective worlds that could allow fundamental distinctions to be made in both. The example of such a principle that I'll describe in the rest of this chapter may come across as excessively outré, but it's at least an instance of the *sort* of concept that may be needed even if not itself the right one. Moreover, as will become apparent in due course, it may turn out to entail ill-understood links with energy manifestations (energy 'eigenstates'), which is a possible 'plus' in the context of SoS theory.

The idea that I'll suggest for a bridging principle grew out of an off-the-cuff remark that Carl Jung made in one of his letters. He said that his 'archetypes' were 'like' the natural numbers. Just as a number can manifest in a vast range of sets of particular objects, so his archetypes could have many manifestations. There are many particular examples of mothers, he remarked, but only a single Mother-idea behind them. By analogy with Jung's analogy, the differences between qualia are 'like' the differences between prime numbers. They are all generically similar but nevertheless irreducible to one another. Moreover prime numbers can combine to generate all natural numbers, just as qualia can combine to produce all possible varieties of conscious experience.

Obviously prime numbers are not promising candidates, so it would appear at first sight, on which to base a theory of quale differentiation, but the same can't be said of prime knots. They share with prime numbers the characteristics of being both irreducible to one another and capable of combination. Knots in general have a proven track record, too, as useful conveyers of meaningful information since the Inca used knotted cords (*quipu*) instead of written dispatches in their communications and for their record keeping. Moreover it's quite conceivable that structures formally equivalent to knots have physical instantiation in the brain, as I shall describe in a moment.

There are vast numbers of irreducibly distinct prime knots; quite sufficient numbers to allow one to suppose that each might have a one to one correspondence with a distinct quale. More than a million prime knots with sixteen crossings exist (actually 1,388,705 such knots). Although there is no general method for calculating the number of primes from crossing number, it goes up very fast indeed – for instance, there are 'only' about a fifth as many primes with fifteen, rather than sixteen, crossings. It's thus likely that the set of prime knots with fewer than eighteen crossings would be more than sufficient to allow a one-to-one correspondence with the entire range of distinguishable qualia.

How could knots manifest in the brain? Well, braids are equivalent to knots and it's entirely possible to suppose that braided patterns of ionic waves or whatever are ubiquitous in brains, just as they can occur in water flowing out of a kitchen tap or cascading down a cataract. The number of crossings in any brain braids would often be quite large enough to allow manifestation of vast numbers of primes. Moreover there are *surfaces* (Seifert surfaces) that are topologically equivalent to knots, including prime knots, which could also well occur in brains in a variety of guises such as bounded ion fluxes or even electromagnetic field patterns. Incidentally Seifert surfaces have to be 'orientable', meaning that any twists in them must be whole turns (i.e. turns of 360 degrees or integer multiples of 360 degrees), not half turns. This restriction offers a way of narrowing the field down if ever it becomes possible to look for them with future brain-imaging techniques. What matters for the moment, however, is that there are ample 'in principle' opportunities for knot-equivalents to manifest in brains.

Knot theory is a somewhat arcane branch of mathematics that crops up in all sorts of surprising areas including quantum field theory. One of the big surprises is that prime numbers and prime knots may turn out to share a common ontological basis. Perhaps it wasn't entirely unrealistic, after all, to look for commonalities between qualia and prime numbers since knots may link the two!

The possible connection of knots with numbers has to do with the Riemann hypothesis and the fact that the positioning of prime numbers on the real number line is related to the occurrence of 'non-trivial' zeros of the Riemann zeta function. The underlying conjecture is that all such non-trivial zeros of the function (which is a complex number function) occur when the real part of complex numbers contributing to a function is $1/2$. The conjecture has yet to be proved, though most mathematicians appear to think it probably correct. For reasons that I'm not competent to understand, it has recently been argued by authors trying to prove the Riemann hypothesis that 'quantum knots' also connect with zeros of the zeta function. Quantum knots are said to be topologically equivalent to manifest knots, in which case the distribution of prime knots should equally relate to Riemann function zeros, thus connecting them with prime numbers.

There's a further twist to the tale due to the fact that quantum knots are each associated with a unique Hamiltonian (energy function). I'm afraid that my argument is going to get even more technical at this stage but there's no way I can express it simply and readers are invited to skip to the next paragraph if they wish. Hamiltonians can be expressed as quantum operators acting either in space or in time. The spatial expression is equivalent to an energy eigenstate, while it can be argued that the temporal equation relates to an SoS. It thus seems possible that the complex Hamiltonians connected with quantum knots may find expression either as complex energy manifestations or as complex SoS manifestations aka qualia.

There seems to be a mathematical paper trail, in other words, that links prime numbers to quantum knots via the Riemann hypothesis and thus to manifest knots (because quantum knots are said to be topologically equivalent to manifest knots). Quantum knots, in turn, may provide a link between objective energy manifestations and hypothetical subjective SoS states, via the Hamiltonians that are implicit in quantum knots and the symmetry breakage that was discussed earlier (a break envisaged as coinciding with energy eigenstate manifestations and representable in terms of the alternative spatial or temporal equations of the Hamiltonian operator). Thus it is just about conceivable that distinctions between qualia share a kind of family relationship with distinctions between prime numbers. Carl Jung would probably have been pleased with the idea, though perhaps his friend Wolfgang Pauli would have dismissed it with the famous comment that was dreaded by so many of his contemporaries when they had explained their ideas: 'not even wrong!'

It's certainly a very shaky paper trail for many reasons, not least that the Riemann hypothesis remains unproven to date, while any connection between braids or Seifert surfaces in the brain and Hamiltonians attributable to quantum knots is far from clear. I've described it for several reasons. First, it offers a conception of the sort of *distinction* between qualia that is required: one that belongs both to the objective world of manifestations and to the subjective world of conscious cognition. There are many sets of three objects gathered together out there in the world, for example, but only one 'number three' – and three, being a prime, can't be decomposed into any other equivalent set. Equally there are a great many examples of blue objects of some particular shade, but only one irreducible 'blue' experience. The correspondence, or analogy if you prefer, between primes and qualia helps to ground qualia differences in a very fundamental sort of reality that is prior to any objective/subjective split, even though the *manifestations* of that reality are split. That's not to say that qualia should be regarded as 'out there in the world' in some sense. Rather, their objective manifestation in the world belongs to the 'objective' aspect of our minds, to the calcium ions or electrical fields or whatever, while their subjective manifestation is as qualia-constituted consciousness.

A second reason for describing the idea is that it also offers a picture of the sort of *entity* that might survive the transition from *extensa* to *cogitans*, from a spatial condition to an essentially non-spatial, albeit possibly temporal, condition. We know that perceptions intrinsic to the material aspects of our brains can and do turn into qualia that reflect realities out there in the world. But a quale belongs to a different category of reality from a pattern of calcium ions or whatever. The concept of a 'qualia space' may well be useful in relation to description of the objective, 'mental' origins of qualia but it's hard to see how the idea could carry over to provide a basis for their subjective differences. However, patterns of calcium ions or other brain events can, in all likelihood, manifest as knots imbued with intrinsic prime-ness and that's a characteristic that does belong equally to the subjective world. Perhaps knots don't actually have this transitive role but, if not, something that similarly has a foot in both camps must surely substitute for them. And knots have what may be considered an advantage over other potential candidates for the role in that they fit neatly with Sherrington's image of a woven tapestry of experience.

A third reason for describing knot theory is perhaps even more suspect; it may be thought too much like an attempt to pile Pelion on Ossa. It's based on the possible connection between knots and energy eigenstates. Since the latter have been envisaged as the flipside of SoS states it would seem that there could be direct translation of aspects of brain behaviour, which as we've seen is a function of history and environment along with factors intrinsic to the brain, into subjective experience. One can suppose, if there is in truth a connection of this sort between knot theory and SoS theory, that brains are devices for directly translating aspects of the spatial topology of the world into the stuff of temporality. Is 'eternity' a word that could be substituted for 'temporality' in this context? So far as I know, there is no way of providing a plausible answer to this question on theoretical grounds. But maybe a look at available evidence will hint at a possible answer. That's where we should go next.

8
ROCKS FROM THE SKY (PART 1)

Back in the eighteenth century natural philosophers were occasionally disturbed by reports that a rock had fallen from the sky. Naturally – in a different sense of the word – almost all of them discounted such fanciful tales for they knew, not from Copernicus or Galileo but from older traditions of belief in the perfection and unchangeability of the heavens, that no such happening was possible. There were no rocks in the sky, so none could fall; meteorites could not exist or, if they did exist, they hadn't come from the heavens.

There is a positive hailstorm available nowadays of reports of 'rocks from the sky' that manifest in a range of guises. Many are incompatible with contemporary mainstream beliefs, both scientific and religious, about our nature. Although they hardly ever figure in issues of prestigious journals such as *Nature* or *Science*, unless to be dismissed, a number of them have been subject to rigorous and sceptical investigation – truly sceptical that is, involving more care than the superficial debunking that is both sadly common and often widely reported. They hold promise of advancing understanding in ways that must almost inevitably elude a purely objective neuroscience. To be consistent, anyone wanting to dismiss them as always valueless must affirm that qualia have no reality, in much the same way that some physicians used to upset their patients by affirming that a pain couldn't be 'real' unless there was visible injury or pathology to cause it. Saying qualia weren't really real was indeed a popular tactic in the late twentieth century but one that's looking ever more threadbare these days.

Whether you can attribute 'reality' to something of course depends on your definition of the term. Many pragmatists explicitly or implicitly reserve 'real' for things that they can see, touch or measure with instruments. As we've seen, contemporary physics supports no such definition since the manifest world is an expression of 'measurements' made on aspects of a realm of potentialities, which itself must be 'real' in some sense despite the fact that unmeasured aspects of it are inaccessible in principle as well as in practice. It's been estimated that around 30 per cent of the gross national product (GNP) of developed countries ultimately derives from applications of quantum theory and thus on the 'reality' of pre-measurement states. And you can't get much realer than what generates a GNP! (Whatever the precise validity of this figure, it's clear that many economically important technologies – lasers, microprocessors etc. – depend on practical applications of non-classical aspects of the world.) It makes more sense nowadays to attribute 'reality' to anything that has effects, whether we know about the effects from direct observation or from well-established theory. Thus, while a unicorn has no reality for it has no existence that could affect anything, the *idea* of a unicorn is real if only because it has resulted in my typing of these words.

Qualia constitute the *only* reality of which we have experience – all else goes on in the dark of unconsciousness – so it would seem perverse to try to deny them that which they *are*. The problem is that we don't yet have any well-established theories about what constitutes their basis or what are their concomitants. To make fundamental progress in any field it's usually best to look at phenomena that *don't* readily fit in with prevailing understandings; it's best to take 'rocks from the sky' seriously, in other words. One needs to develop tentative ideas that might possibly provide some sort of basis for explaining the occurrence of especially awkward facts and then try to work out whether the ideas do actually fit these facts or whether they need to be modified or abandoned. It's a process at the basis of all science that Imre Lakatos described, probably better than anyone else has done since, in his *Conjectures and Refutations*. He was an interesting thinker who had a remarkable, but sadly brief, life (he died aged only 51) having been a Communist Party functionary in Hungary who escaped to London where he became a student of Karl Popper's. If he had lived longer we should probably all now have a better understanding of the nature of science than is readily available to most of us.

The first half of this book outlined notions of a type likely to be needed if we're ever to understand our own nature. From a purely material, objective point of view, we're plastic beings moulded by our histories who resonate, so to speak, with our environments. That aspect of us is adequately covered by biology, the neurosciences, psychology, sociology and the like. They've still got a long way to go before reaching any complete description of our objective aspects but it's unlikely that any radical change of direction will be needed. It should prove especially informative to have more and better calcium imaging available according to the ideas offered earlier, as far as neuroscience is concerned for example, but that will almost certainly happen in the normal course of events.

The most important part of our nature however, what makes us *us*, lies in our consciousness that connects with the material side of our nature somehow but that seems to belong to a primarily non-spatial aspect of reality. Are any of the 'rocks' that exist out there likely to relate to ideas described in Chapters 6 and 7 and can they help with refining the conjectures? That's the main theme of this chapter and the next two; then we'll move on to see whether any predictions can be made about the likely existence and nature of currently undiscovered types of 'rock', followed by a look at some of the potential implications of the whole picture.

An immediate problem for this agenda is that there's so *much* potential evidence. Better too much than too little perhaps, but the embarrassment of riches is bound to make for difficulty when it comes to distinguishing what's possibly relevant from what is probably irrelevant. The first step needed, therefore, is to separate potentially helpful awkward facts from ones likely to prove unhelpful. By the way I'll only be discussing phenomena that have been shown beyond reasonable doubt to pose awkward problems, not ones likely to prove susceptible to mundane explanations. The status of such phenomena isn't beyond *un*reasonable doubt, of course, but there are plenty of excellent, up-to-date surveys and discussions of them available nowadays (e.g. books by Julia Assante, Chris Carter, Dean Radin, Greg Taylor and Michael Tymm, among many others) that should convince any but the most prejudiced of their deeply puzzling nature.

The founders of the SPR were interested in five principal topics: hypnotism and phenomena associated with it; telepathy (a term coined by Frederic Myers who was among the SPR founders); apparitions; physical phenomena associated with poltergeists and mediums; and, finally, 'spirit communications' relayed by mediums. It's worth considering some of these topics and subsequent developments relating to them before moving on to newer foci of interest. Which of them, if any, are likely to help with our conjectures and refutations?

The Victorians were especially interested in hypnotism because it was the child of mesmerism, offering hopes of cure, lessons in subjection and a route to achieving apparently paranormal powers of telepathy and clairvoyance. That's far too much of a hotchpotch to be useful for our purposes. It would involve us in considerations ranging from psychosomatic medicine to Ganzfield techniques that are best dealt with separately, if at all. Indeed, although the nature of hypnosis still isn't properly understood, it seems quite likely that it has primarily to do with unconscious mind, having only secondary consequences for conscious experience. Given the view of unconscious mind developed earlier we could picture it, in a loose sort of way, as involving extensive landscape reconfigurations negotiated via external inputs; puzzling indeed, but not necessarily posing any great 'in principle' problems as to its origins despite the oddity of some of the phenomena associated with it.

Telepathy, however, is a definite 'rock from the sky' albeit a somewhat lightweight one. The evidence that it occurs is overwhelming, but with an 'effect size' that's probably quite a bit smaller even than the 'pre-sponse' one (of 0.2) mentioned in Chapter 4 – though that's only a guess as I don't know of any formal calculation of effect size made on the totality of telepathy findings in the way that has been done for pre-sponse results. Telepathy, too, seems to relate more to 'mind' than to consciousness per se, although the *how* of its connection with mind remains a mystery. Some recent studies have shown that two people's individual EEG activities can sometimes synchronise when they are thinking of one another or feel 'in sympathy' with one another, even when they are placed in separate rooms with acoustic, visual and electromagnetic shielding. Maybe EEG synchronisation between individuals could provide some sort of basis for the occurrence of shared awarenesses, though of course it only shifts the mystery to the question of why EEGs should harmonise. And it could equally be the case that the shared awareness causes the EEG changes rather than vice versa. One might speculate that consciousness has a part to play after all since the subjects probably do have to be consciously aware of their 'in sympathy' thoughts for the effect to occur. Nevertheless it all seems a bit too nebulous to be useful in our enquiries.

Apparitions are the next topic that interested the Victorians. Edmund Gurney gathered accounts of over 700 cases that were subsequently added to, collated and reported in *Phantasms of the Living*. Can we say anything more definite about their origins today than he was able to suggest back in the early 1880s? The answer is probably 'yes', thanks mainly to a case report that is not nearly as well-known as it deserves. It was published in book form in 1980 (*The Story of Ruth*) and seems to have been largely forgotten subsequently. Case reports in medicine have often provided vital clues to the aetiology of obscure phenomena and are rarely dismissed as 'mere anecdotes' in the way that can happen in other fields. This particular report was written by psychiatrist Morton Schatzman about one of his patients, though she turned out not to have any diagnosable mental illness. What she did have was an extraordinary talent for seeing apparitions at will, something that had naturally enough frightened her before she understood her own responsibility for their occurrence. They appeared entirely realistic and natural, she said, and a lot more than simply 'seeing' was involved since she could also hold conversations with them that sometimes elicited personal and other information that she had not been consciously aware of remembering.

It's clear that her apparitions had no 'objective' basis that was independent of her own mental processes since, following a bit of practice, she was able to generate ones of *herself* at any apparent age from infancy onwards, and of Morton Schatzman while he was in the same room as her. But they did have 'objective' correlates. An especially striking finding was that when she 'combined', as she put it, her actual self with an apparition of her pre-reading-age child self, her performance on the Stroop test went from that of a normal, literate adult to that expected of a pre-literate child. And there was no way she could have consciously simulated a change like that even if she had known about Stroop tests, which involve subtle changes in reaction times dependent on ability or inability to read the meaning of words such as 'blue' or 'red'. Electrophysiological studies done when she was experiencing apparitions showed changes in evoked potentials in her brain consistent with visual perception of something real, but not the retinal activity that would normally be expected.

Although unique, so far as I know, in having been studied so thoroughly, 'Ruth's' (not her real name; Schatzman was scrupulous about preserving her anonymity) case shows that she was capable of generating subjective apparitions that appeared to her objectively real in all respects. Presumably we all have a similar ability to some extent, even if we never use it. She was or is (she's probably still alive) an amazing performer; a Usain Bolt of the perceptual world, it seems. But nearly all of us can traverse a hundred metres even if we don't do it nearly as quickly as him. It's likely, therefore, that apparitions are best regarded as subjective constructs; they are hallucinations unassociated with delirium or mental illness, in other words.

This still leaves questions about where the motivation and information to make an apparition might come from. In 'Ruth's' case the original motivation for her rather obsessive iteration of apparitions was probably down to childhood unhappiness linked to an eidetic memory – when she was expressing her child self, she showed eidetic memory on testing that was no longer there when she was being her normal adult self. Subsequent motivation was provided mainly by Morton Schatzman who encouraged her to exercise her remarkable talent in order to gain control of it. The necessary information presumably came from her personal memories, many of them held unconsciously. Most of Gurney's cases however, like most of those subsequently reported, were one-off occurrences. Some were apparitions of friends or relatives who appeared to convey information that was both unknown and unavailable to the experiencer at the time the apparition manifested – for example, announcing that the person 'seen' had just died. Very occasionally two or more people reported experience of the same apparition at the same time.

While a few occurrences of information bearing apparitions might be put down to chance, that's an unconvincing explanation for most of the reports in view especially of their frequently one-off nature. Others, including ones that don't come bearing information, may be produced by what might loosely be termed auto-suggestion. Nevertheless it does looks as though both the motivation and information needed to produce an apparition may occasionally have paranormal origins, even if the apparition itself doesn't. Apparitions conveying veridical information previously unknown to a perceiver are perhaps best viewed as an unusual expression of a telepathic cognition. Even if there were any reliable way of definitely identifying apparitions with a paranormal contribution to their origins, therefore, it seems that they should probably be regarded as being at least as unlikely as straightforward (!) telepathy to take us much further.

Physical phenomena look much more promising from our point of view, despite being both rare and hard to believe. They raise obvious questions to do with energy conservation, for example, that may be useful when it comes to developing ideas about possible 'back action' of SoS systems on material ones. Anyone who thinks they can all be readily explained away in terms of fraud or self-deception needs to read psychologist David Fontana's account of the Cardiff poltergeist that he investigated or philosopher Stephen Braude on the apparently unsophisticated 'gold leaf lady' who could exude flecks of brass foil from her skin, something that skilled magicians were unable to match or to explain.

Other than mysterious and apparently rather random happenings such as those above, where it's anyone's guess as to what might have been going on, there are two very different varieties to think about: first, the physical phenomena reported in séances – rappings, table tippings and 'materialisations' were said to be the most frequent in Victorian times; second, the alleged effect of conscious intention on supposedly random systems.

Séance happenings were certainly often due to conscious or unconscious fraud, but some weren't. Daniel Home, for instance, who reliably produced or enabled a range of remarkable phenomena during his sessions, was rigorously investigated by people with wide experience of fraudulent mediums but was never caught out. Given the care taken, it's almost certain there was nothing *to* catch out. Similarly the 'Scole group', who were active in the 1990s and specialised particularly in producing remarkable light effects during their sessions, passed careful scrutiny by experienced investigators as far as many of the phenomena were concerned, though doubt remained about some. There are very valid questions to be asked about who produced the phenomena and how they did it.

Participants in séances generally attribute the 'who' to 'spirits', but it's far from clear that this is necessary. The conditions of a séance, the heightened anticipations in a group of people with similar interests assembled in dim light and often holding one another's hands, provides a recipe for the emergence of a 'group mind' like the one that the Mormons experienced (see Chapter 2). There need be nothing paranormal about the development of a group mentality of this sort; it is a natural consequence of a shared group dynamic. Maybe it's the group mind of the participants that is somehow responsible for the phenomena. That would certainly go a long way towards explaining why inclusion of a sceptic in the group can result in a 'no show' – an outcome that the sceptics themselves often misinterpret as proof of their case.

Can the apparent phenomena be regarded as a group hallucination then, and no more objectively real than were 'Ruth's' apparitions? Did the phenomena have only subjective, not objective, reality in other words? It's a tempting hypothesis, which may meet some of the facts but struggles to account for others such as physical effects found to have persisted after a group had dispersed. It does look as though something owning conscious mentality, a 'spirit' or more likely a group mind, or perhaps even both together on occasion, can indeed directly produce genuine and often meaningful objective physical effects. Which brings us back to the apparent violation of energy conservation; something that most physicists would say can't happen.

But maybe it can. We noted earlier that energy conservation is a consequence, via Noether's theorem, of the indifference of physics to displacements in time, i.e. in *clock* time. We've proposed that consciousness is a concomitant of a different sort of time, i.e. the succession of present moments embodied in SoSs. Noether's theorem applies only to *smooth* transitions but the relationship between clock time and SoS time is not smooth. Thus energy conservation would be unlikely to survive any displacement involving a transition between clock time and SoS time. We therefore have the outline of a hint that, if SoS systems can have any 'back action' on physical ones, considerations of energy conservation need not necessarily apply.

The other type of well-studied physical phenomena, to do with effects of conscious volitions on physical systems, doesn't imply any violation of energy conservation but raises rather different questions and issues. The most thorough tests ever performed on this were carried out by Robert Jahn, Brenda Dunne and colleagues over a twenty-eight-year period at Princeton (Princeton Engineering Anomalies Research – PEAR). They finally gave up when they felt that they had proved their case far beyond all reasonable doubt and there was no point in further testing. The basic idea, conducted in a range of variations, was to provide people with a measure of the frequency of random physical events, usually radioactive decays, and then ask them to form a conscious intent to make the measure show either an increase or a decrease in the rates of the events. The measure, of course, provided an accurate objective representation of the actual clock-time rates.

To cut a long story short, an effect of volition on rates was found with accumulated odds of trillions to one against it being due to chance. Moreover subjects were unable to influence rates of pseudo-random events (i.e. deterministic events generated by computer programs that mimic truly random outputs), suggesting that the effect wasn't due to any direct 'psycho-kinetic' influence on the rate measuring apparatus. People really were influencing events that should have been random. However, the effect was extremely weak. Voluntary intent appeared to bias only about one in ten thousand events overall. No-one was going to beat the odds in a casino via the effect, which only showed up basically because PEAR allowed vast numbers of 'casino throws' to occur every minute.

Nevertheless the effect was real. How can it be understood? Truly random events are down to the randomisation that occurs along with quantum measurement outcomes; what have been called by physicist Henry Stapp 'Dirac choices'. Anything that influences these 'choices' must do so via an effect on the pre-measurement state, i.e. the wave function. The PEAR researchers have offered formal, mathematical descriptions of how this might occur but I think it's fair to say that these are a little too abstract to provide much insight into what might have been going on.

One of the most surprising characteristics of the effect is that it was meaning-dependent. It wasn't down to some informational 'difference that makes a difference' only. If people intended rates to increase they generally did; if they intended them to decrease that was what mostly happened. It wasn't just a case of any old contact of mind with quantum system having an effect on the system. Increase and decrease are, of course, meaningful concepts and meaningfulness inevitably eludes the type of formal description offered by the PEAR team.

Put in terms of the metaphor offered in this book, one might say that the 'valleys' of the landscape of extended mind have been shown capable of extension into the pre-measurement world of potentialities, influencing measurement outcomes in meaning-related ways. This is very surprising since the 'landscapes' were envisaged as images of a purely classical, post-measurement dynamic. There's no way, one might think, that a classical dynamic can reach back into any aspect of the pre-measurement states that contribute to it. The arrow of time (causes precede effects) prevents any such happening, doesn't it? SoS landscape equivalents (i.e. mappings from objective brain energetics into SoS format), however, would not be subject to this restriction because they bridge successive instants of clock time with their extended present moments. They might enable apparent causal reversal across the 'measurement' boundary if they extend out into their environments as do the 'objective' 'landscapes' that they model. An effect like that could perhaps manifest as an appearance of back action on objective systems.

It remains all very much of a puzzle, though, especially because the 'landscape' metaphor doesn't work at all well if applied to notional SoS mappings since these are viewed as being represented in a primarily temporal format, not in the sort of spatial format that a 'landscape' image evokes. But at least having to abandon the spatial metaphor makes it a bit easier to see how SoS forms *might* interact with the world of pre-measurement potentialities since that too is basically non-spatial, at least as far as entanglement phenomena are concerned. Just possibly some primarily atemporal, in the sense of an influence independent of clock time, effect is involved in producing the anomalous findings. If so, the PEAR findings may be a different manifestation of the same circumstance that allows apparent violation of energy conservation.

To sum up, two varieties of physical paranormal phenomena whose occurrence has been established quite as firmly as that of many widely accepted non-paranormal phenomena, provide support for the view that conscious mind can affect objective physical systems both directly and via a (very weak) ability to bias quantum measurement outcomes. However, both types of effect appear to violate accepted physical principles (energy conservation and causal directionality respectively). The violations can be explained, in principle at least, if conscious mind has temporal properties inherently different from those of the 'objective' world, properties of the type suggested in Chapter 6.

I'll leave the final Victorian preoccupation, that with supposed 'spirit communications', for the next chapter because I need to slot it in with other, more recent developments.

9
ROCKS FROM THE SKY (PART 2)

Back in 1871 an up-and-coming Swiss scientist named Albert Heim, who subsequently became quite famous for his contributions to geology, fell off an alp. He was aged twenty-two at the time and came to a halt on a ledge about seventy feet down, not seriously injured. He later recalled experiences as he was falling:

> Mental activity became enormous, rising to a hundred-fold velocity or intensity. The relationships of events and their probable outcomes were overviewed with objective clarity. No confusion entered at all. Time became greatly expanded ... there followed a sudden review of ... [my] entire past ....

So impressed was he by his experiences that he subsequently chased up and interviewed other survivors of potentially fatal accidents, publishing an account of their stories in 1892. Many had told of experiences similar to his own. Evidently he was describing something closely resembling the 'life review' reported by around 5 per cent of modern near-death experience (NDE) cases – on which more later. Unlike most NDE cases, however, his experience occurred in the context of anticipated death while he was still falling, not in that of the actual cardiac arrest or whatever more usually associated with NDEs.

These 'life reviews' raise lots of interesting questions in relation to SoS theory, but I want to focus on just one such question at present. We've pictured subjective experience as being like a sort of braided rope assembled from overlapping moments of temporal experience (i.e. individual SoSs), woven and stretched out over the course of clock time by the energetic objective events with which they are associated. What might happen when the 'weaving' process finishes or is anticipated to be about to end? Remember that the rope's component SoSs *are* moments of subjective time, so they're not themselves going to be snuffed out when the weaving process stops in clock time. The very basis for the existence of their time, following the symmetry break envisaged in Chapter 6, is independent of clock time's basis; the one relates to a 'subjective' world, the other to an 'objective' world.

One possible outcome of coming to the end of the 'weave' is that the 'rope' of experience would fall apart and all its little components drift off separately into the empyrean realm of 'subjectivity', which is to be considered as real as, though separate from, the realm of objectivity. The other possibility is that the 'glue' of temporal overlap is strong enough to resist separation. No longer stretched out along clock time, experience might change its topology from linear to something more like a sphere. A change such as that would certainly fit reports made by people that they had experienced their whole past all at once. But is there any evidence that assemblages of this sort do in fact stay 'glued' together and able to persist from the perspective of our own clock-time, objective lives? Any such assemblages wouldn't be much like people as we ordinarily conceive of them, but they would incorporate person-like memories and perhaps other characteristics. There are two main lines of evidence to consider, both of them surprisingly good in some respects albeit still firmly in the 'rocks from the sky' category from the point of view of current scientific orthodoxy.

The first is to do with evidence for 'reincarnation'. Erlendur Haraldsson, an Icelandic professor of psychology, and Ian Stevenson, an American psychiatrist, collected an impressive amount of evidence from children that something of the sort occurs. Although Haraldsson retired a while ago and Stevenson has now died, their colleague Jim Tucker, another psychiatrist, continues to investigate new cases to the same high standards, while a number of other people have embarked on similar investigations. Results so far show that reincarnation is more often reported in societies that believe in it (mainly Indian), but cases have cropped up in all major cultures. The child generally starts talking about memories of 'another family and another life' soon after it learns to speak. If no great attention is paid to them, the memories fade and are more or less gone by the age of eight or so. When the child has provided enough detail, it has sometimes been possible to chase up the 'other family' who have generally provided confirmation that some dead relative of theirs matched the child's memories of a 'previous life'. A few alleged previous incarnations have also been identified from independent historical records, using information provided by a child. Children themselves have spontaneously recognised people and objects from their 'old' life without any prompting, despite having had no possibility of gaining information about them previously in 'this' life.

Interestingly enough, far more than would be expected by chance of 'old life' deaths were due to violence. Sometimes the child has a birthmark corresponding to the cause of death if it was due to a gunshot wound or other localised trauma. Stevenson also looked for evidence of 'karma' as some of the alleged previous personalities were unpleasant, as might be expected given their high death rate from violence. He didn't find any evidence for it though, in that the social circumstances of children appeared to be distributed randomly with no sign of any relationship between the harshness or ease of a child's circumstances and the nastiness or otherwise of the alleged previous personality.

Of course talk of 'previous lives', along with so-called 'past-life regression therapy', has become popular in some circles in the West perhaps as a consequence of widespread interest in Buddhism. The difficulty is that most of the 'evidence' for it in adults derives from hypnotic sessions. The 'alien abduction' phenomenon (Chapter 2) and many similar fads show that anything hypnosis-associated has to be taken with very large pinches of salt. There's no known way to readily and directly distinguish hypnotically induced false memories from veridical ones. Investigators have had to rely on indirect means, which are troublesome enough when it comes to court cases involving this-life memories that may or may not be false, such as memories of sexual abuse in childhood or adolescence, for example. Distinguishing true from false is even more suspect and difficult when it comes to alleged past-life memories. There are rare and controversial accounts, such as the 'Bridey Murphy' case, that suggest that memories of a past life correspond to historical events unknown when the memory was reported, but anything approaching proof beyond reasonable doubt is very elusive indeed.

It does appear, to be fair, that descriptions people give of their alleged past-life circumstances indicate occupational and social class distributions typical of those existing at the alleged time of the previous life. In other words, the descriptions given are of a thousand rural housewives or farm labourers for every duchess or hero. The popular notion that most such claims are of having been Cleopatra or Alexander the Great is not true. Nevertheless evidence from past-life regression is too nebulous and shaky, too contaminated by ifs and buts, for confidence that it can tell us much. The evidence from the spontaneous claims of children is much better and well worth taking seriously.

And now for the second line of evidence – the jewels that lurk in the murky waters of mediumship. Jewels are rare indeed, but they do exist amid the wishful thinkers, self-deceivers, outright charlatans and feeble talents attracted to this field. Mrs Leonora Piper and Mrs Gladys Leonard are the examples most often paraded. Both were subject to rigorous investigation – more rigorous on occasion than would be permissible nowadays on ethical grounds – to exclude any possibility that they used hot reading, cold reading, fishing techniques or any other of the tricks beloved of conscious or unconscious frauds. Yet they continued to provide information in their sittings that couldn't have come from any normal source. The Brazilian medium Chico Xavier (1910–2002) was perhaps the brightest star of them all. He completed primary education only but eventually got a clerical job in a Ministry of Agriculture branch office, which he held for most of his life. His major talent was for automatic writing, poems as well as prose, which resulted in his publishing 458 books in total said to be in the varied styles of a range of well-known but dead Portuguese and Brazilian authors. He was seen on occasion to be writing two different books simultaneously, one with each hand at great speed, allegedly to the 'dictation' of the real, albeit deceased, authors. The books earned him quite a lot of money, but he gave it all away and lived on the modest salary provided by his day job.

However, it appears that some talent for mediumship isn't very rare. As with most faculties there are a few outstanding performers and a lot more relatively undistinguished ones. The Windbridge Institute in Arizona has actually run courses for would-be mediums and certified those that made the grade – around twenty so far, I believe. Typical mediums are said to be more often female than male and to have more than their fair share of physical illnesses, but otherwise rarely fit the fey stereotype so often portrayed in films and books. Some of the most able (e.g. Winifred Coombe Tennant whose pseudonym for mediumistic purposes was 'Mrs Willett', and Geraldine Cummins) have been noted for their robust common-sense in everyday life.

What can be inferred from mediumship phenomena? It seems that mediums can intermittently (even the most reliable have their off days) contact entities that fit the notion of an 'SoS assemblage'. They do so in various ways. Some, especially the automatic writers, appear to allow an entity use of their brains for a time. In Chico Xavier's case, when he was writing different books with either hand, it looked as though two entities were simultaneously using opposite sides of his brain! It's a phenomenon reminiscent of hypnotic automatisms with the big difference that the hypnotiser is both invisible and sometimes able to provide information unknown to the medium or his/her associates. Other mediums hold what appear to be conversations with entities, though they are conversations in which both sides are more than a bit deaf and sometimes have to relay information through third parties (i.e. a medium's so-called 'controls'). The phenomena certainly provide further evidence that apparently independent SoS assemblages can influence physical systems (i.e. a medium's brain), though whether they do so directly or via the medium's own intrinsic subjectivity, or both, remain open questions.

Given that SoSs are envisaged as existing in a different sort of time from that of the objective world, it might be supposed that the medium is reaching back somehow, from his/her objective perspective, to the consciousness of people dead from that perspective but perhaps still incarnate from their own subjective perspective. There's quite a bit of evidence, though, that this is unlikely to be true since some at least of the supposed SoS assemblages seem to have changed and developed since they were incarnate. Xavier, for instance, didn't reproduce books already written by his alleged authors, but new ones in their style.

Perhaps the best available evidence for post-mortem change was provided by the 'cross correspondences' case in which entities claiming to have been founding members of the SPR themselves suggested an 'experiment' in which they would provide partial references to some piece of arcane knowledge in separate communications to different mediums. Only when the separate references were combined, the entities claimed, would their meaning become clear. These communications continued for some thirty years and involved a medium living in Boston (Leonora Piper), one in India ('Mrs Holland', who turned out to be Rudyard Kipling's sister) as well as several living in different parts of England. Some correspondences of the type that the communicators 'said' they intended were identified by researchers (principally John Piddington). These seem very unlikely to have been down to chance but unfortunately the 'experiment' didn't lend itself to any formal statistical treatment, so can and has been dismissed on those grounds. Nevertheless the fact that alleged communicators 'suggested' such a novel idea is itself significant and can be taken to indicate that, whatever their true nature may be, it encompasses more than a fixed and unalterable memory.

Combining evidence from childhood 'reincarnation' memories and aspects of mediumship provides quite a strong case for supposing that entities referred to as 'SoS assemblages', or personal subjective awareness if you like, can persist independently of any objective incarnation and can continue to influence physical systems, i.e. the brains of children and mediums and perhaps the bodies of children too, if the birthmark evidence is to be believed. Whether this influence is of the same sort as that directly producing the 'physical phenomena' discussed in the last chapter isn't clear. It could be indirect and mediated via the consciousness of the child or medium.

There's one other 'rock from the sky' that I should mention before moving on to 'verge of death' experiences, which are now widely studied and beginning to lose their former 'rock from the sky' status. The remaining awkward fact is that of clairvoyance, which is a bit like telepathy. Indeed it's often not clear whether some particular paranormal event should be attributed to clairvoyance or to telepathy. SoS theory muddies these waters even further with its separation of subjective from objective time, since it would allow many cases of apparent clairvoyance to be attributed to trans-(objective)temporal telepathy and sometimes might allow the reverse attribution.

If telepathy isn't likely to provide useful evidence relating to SoS theory, the same is going to be true of clairvoyance. It does offer one intriguing hint, though. The American Stargate Project (run from 1978 to 1995) was aimed at ascertaining whether clairvoyance is good for spying; whether remote viewing could provide useful intelligence, as the official euphemism had it. The overall conclusion was that it couldn't because it was too unreliable, which might have been predicted from the 'effect sizes' already known about from parapsychology work existing at the time the project was started. It's said there was a scare that the Russians were working on the same approach so the Americans had to see whether 'effect sizes' could be improved. The answer was that they couldn't, or not sufficiently for practical purposes. Ganzfield techniques, for example, produced small improvements but nothing large enough to provide information that could be relied on, nor was any means discovered of differentiating veridical from illusory clairvoyant reports.

However, the Stargate Project did identify a few star performers able to produce remarkable, albeit occasional, 'hits'. Apparently these people were often best at remotely viewing installations with a high-energy usage. According to SoS theory high-energy use is going to be associated with a high rate of generation of very ephemeral and mostly unco-ordinated

SoSs. Nevertheless, as these belong in the same ballpark as a would-be remote viewer's 'subjectivity', maybe they produce a sort of searchlight effect that allows their 'objective' origins to be remote viewed more readily. As I said, it's only a hint that SoS theory may be on the right lines and perhaps not a very good one! Let's move on to see what NDEs and the like can tell us.

10
LIFE AT THE EDGE

What we have so far is a theory recasting our conscious minds in terms of interwoven or interlinked threads of tiny experiential chunks of time (SoSs), entities that are informed by the structure of unconscious mind and stretched out in clock time. Sources of evidence that are generally, but sometimes quite unjustly, regarded as unreliable and flaky indicate that conscious minds can influence objective physical systems to a limited extent and that such minds do not necessarily fall apart or disperse when the physical processes that have 'woven' them are disrupted, whether temporarily (in clock-time terms) or permanently. Any influence of conscious mind on physical systems, so the theory suggests, is not necessarily constrained by either energy conservation or (clock-)time's 'arrow'. The theory also allows for a possibility that, when conscious mind detaches from the structured energetic processes that shape it, its overall topology may switch from linearity to some other form that might loosely be pictured as spherical. There is limited evidence, too, that its new form is not necessarily fixed and that it can continue to appear to evolve and change, at least from our clock-time perspective.

Studies of NDEs and, to a lesser extent, of death-associated experiences are becoming almost mainstream nowadays and there's a large, ever-growing literature on them. They're particularly valuable from an evidential point of view when they are prospective (e.g. the NDE studies conducted by Pym van Lommel, Sam Parnia and Penny Sartori) rather than retrospective. It turns out that up to 20 per cent of people revived after a cardiac arrest can report experiences with at least a few of the characteristics of the stereotypical NDE that has become almost a cultural cliché: experiences of finding oneself out of one's body; entering a tunnel that leads to the suburbs of a paradise populated by deceased significant others; life review and so forth. Cases reporting most or all of these features are much less common, but do occur. Similar reports also occur, but are less frequent, following other causes of actual temporary death or even just anticipated death as in Albert Heim's account (see Chapter 9).

There's an almost equally large literature detailing attempts to explain NDEs in terms of contemporary neuroscience, neural 'emergentist' views of the nature of consciousness, or even as cultural constructs. It would be tedious indeed to discuss them at length. However, it's worth commenting that the 'cultural construct' view makes no sense whatsoever since NDEs have been reported, with only minor variations, from all cultures and all periods of history of which we have records. They've also been reported by small children who couldn't possibly have learned anything about the experience before having it.

Neural emergentist theories of consciousness struggle to cope with well-attested observations that people have reported veridical perceptions while 'out of their body'; they've described actual events that occurred at times when their EEGs were known to have flat lined, along with other out-of-this-world experiences. The most dramatic and often quoted of these cases was the 'Pam Reynolds' one. She described happenings in the operating theatre, among other experiences, that occurred after she had been anaesthetised but probably shortly before she was put into a totally inert, hypothermic and exsanguinated, state when her blood circulation stopped during a period of induced hypothermia for repair of an intracranial aneurysm. Neural activity in her brain, and of course associated energy production, must almost have ceased at the *apparent* (clock) time that she was having her out-of-this-world experiences, although there may have still have been some activity while she was having the reportedly earlier perceptions of the operating theatre. There is room for argument in her case about whether the subjectively reported timing of her experiences coincided with the clock time of events during her operation. Other cases, while less dramatic than hers, provide rather better evidence that clear and detailed, veridical perceptions can be recalled of events occurring at clock times when a patient's brain couldn't possibly have been functioning anything like normally.

Attempted neuroscience explanations in terms of oxygen deprivation, endorphin release, temporal lobe micro-seizures or whatever all have their problems. Oxygen deprivation is a non-starter as an explanation for a whole lot of reasons, not least that NDEs have been reported by people who were never oxygen deprived at any time. There is overlap between NDE symptomatology and that of some psychedelic drug trips, and there is a theoretical possibility that a potent endogenous psychedelic (DMT) might accumulate in traumatised brains; equally endorphin release might theoretically account for some of the feelings of peace and so forth reported during NDEs. Despite claims to the contrary, NDEs have little in common with symptoms of temporal lobe disturbance. In any case, supposed causes of this sort would have to be envisaged as operative at times when *no* experiences or feelings of *any* sort would be expected. Why should people later recall induced experiences that they couldn't possibly have actually had according to any simple 'neuromodulator release causes it' or similar view? The most that one can say is that 'neuroscience' factors might contribute to some of the symptomatology of NDEs, but such factors almost certainly can't be used to explain them in the way that so many late twentieth-century authors hoped to achieve.

SoS theory, on the other hand, does offer concepts of a type that seems to be needed in order to begin to account for some at least of the phenomena. For instance, the change in topology that may occur when conscious mind is no longer 'stretched out' along clock time might, as mentioned earlier, provide a basis on which to explain some aspects of the 'life review' experience and perhaps also the '360-degree vision' that some NDEers have reported. Of course the experience of 'seeing' any aspects of the objective world, such as goings-on in the operating theatre while the patient was having their NDE, is hard to explain. It is even harder to explain if one believes the reports of a few people blind from birth who are said to have experienced 'vision' for the first time during their NDE ('Grey Mary', eat your heart out! – see Chapter 5). And there's no good reason for *not* believing them, so far as I know.

One might expect any NDE 'vision' experience to be entirely subjective and derived from memories of a lifetime of visual perceiving. Perhaps, as some have suggested, sensory leakage can provide a channel allowing access to the objective world during an NDE, but that couldn't possibly account for anyone experiencing vision for the first time. In any case NDEers generally say that their vision, and other aspects of their experience, is unusually vivid and intense. Sensory leakage may sometimes contribute something to NDE experience, perhaps, but can't explain it any more than endorphin release or the like can explain it. Could an SoS assembly topology change account for any of this? Well, it might just allow 'vision' to the previously sightless by enabling a synaesthetic linkage from other types of perceptual memory. But how could these altered topologies encompass experience of objective events in operating theatres and the like?

This is not an easily answered question! As a first step, it has to be remembered that the change is envisaged as one in the structure of subjective *time*. And this raises issues that may be relevant to another and perhaps the greatest NDE puzzle, or at least the greatest approachable puzzle. Getting to grips with that may help towards forming some idea of how subjective entities might directly perceive objective ones.

I guess we'd all like to know whether the 80 per cent or more of people who don't report NDEs haven't had them or have simply forgotten them. That's a question impossible in principle to answer, but we can usefully ask how on earth some people *are* able to remember them. This is a problem for conventional and SoS theory alike. Conventional theories have to explain how brains lacking visible neural activity can form memories of unusually vivid and detailed experiences. SoS theory faces the problem of explaining how experience happening while 'detached' from a brain can later be recalled so vividly via, presumably, neural mechanisms.

Conventional theories have to postulate that both the near-death experiences and the memories of them occur and gel at the time of waking or shortly before. This is plausible from one point of view in that there is the close linkage between consciousness and early stages of memory processes described in earlier chapters. It's less plausible in that people waking up in these circumstances are generally woozy and confused, far from able to have remarkably vivid and detailed experiences of any sort whether veridical or hallucinatory. And, of course, the theories struggle to account for veridical perceptions reported as having occurred while an NDEer perceived himself or herself as being 'out of her body'. Proponents of this sort of explanation generally have to deny the truth of such reports, which does rather run counter to some of the available evidence.

SoS theory, on the other hand, suggests that experiential time gets detached from clock time during an NDE and assumes a different 'shape'. When experience rejoins 'its' brain, it must resume clock-time linearity and get extended into a form that might be recallable *provided* it is capable of back action on physical systems. Thus, the 'conventional' idea that experiences and memories gel on waking may actually represent what happens from an objective viewpoint. The conventional idea may be entirely correct from a neuroscience point of view, in other words, but doesn't and can't represent the truth of subjective temporality.

This proposal doesn't obviously account for how an experiential entity, detached from its brain, could gain information about happenings in the objective world but it does allow the beginnings of a conceivable explanation. Detached from clock time in the manner envisaged, all bets are off as to the capabilities of a conscious mind. Its capacity for clairvoyance, for instance, might be greatly enhanced and allow direct perception of the objective world while in its altered form. It's a suggestion that could be put in the 'clutching at straws' category were it not for the fact that some NDEers have claimed new or enhanced paranormal faculties following their experience. I don't think these claims have ever been formally tested, although it is known that people undergo personality changes – often quite profound changes in the direction of being less materialistic and more charitable – after their experience. There does seem to be some rather weak evidence consistent with a possibility that altered capacities during an NDE might have a sort of 'hangover' effect that could reflect greater abilities, perhaps including clairvoyance, that were present during the experience.

Probably the biggest conceptual difficulty with this line of thought is to do with trying to envisage how an SoS entity detached from clock time could lay down the neural memories needed for its recall on re-joining its brain. Is there any evidence that conscious minds in the form envisaged can affect the objective world so profoundly and in such exquisite detail? Psycho-kinetic effects of the type discussed in Chapter 8 are surely far too weak and unreliable to do the job. Nevertheless evidence does exist pointing to effects with the necessary specificity, comprehensiveness and precision. Indeed the Chico Xavier type of automatic writing and perhaps other mediumistic phenomena suggest that minds of some sort, ones that apparently possess a degree of independence from a brain influenced by them, can impose complex and elaborate behaviour on the influenced brain. Perhaps it isn't too unreasonable, after all, to make the assumption that SoS assemblages can inform neural behaviour in the way needed for recall of NDEs.

Additional evidence that such effects can and do occur relates to so-called 'end-of-life' experiences and happenings. All sorts of apparitions, telepathic cognitions and synchronistic physical happenings (clocks stopping, radios switching themselves on, etc.) are reported by the bereaved and those about to be bereaved. But I don't want to discuss those because, from an evidential point of view, they are probably no more useful than the telepathy and apparition events mentioned previously. Moreover there are so many deaths (nearly 100 per hour on average in the UK alone) that worries about chance and coincidence have to be taken into account, which confuse the issues even further.

I want to focus instead on reports provided by hospice nurses and other professional carers for the dying, who have less emotional involvement in outcomes and less of a 'large number' problem when it comes to questions of coincidence. Accounts have been systematically gathered from them (e.g. by Peter and Elizabeth Fenwick). Carers consistently tell of cases where a dying person is heard to hold 'conversations' with deceased friends or relatives who have sometimes even provided accurate information about subsequent time of death – or so the person has told his or her carer(s).

More remarkable still are apparently quite frequent cases of 'terminal lucidity' when a confused or demented person suddenly regains lucidity or 'becomes their old self' for a short period before dying. It's almost inconceivable that terminal lucidity in advanced Alzheimer's cases, for instance, could be down to their neurophysiology for they no longer possess the connections, circuits and neurons that supported their 'old self'. But it is perhaps conceivable that a topologically altered conscious mind would harbour the appropriate representations of self and personality, all bound up in one contemporary package from a clock-time perspective. Maybe conscious mind can begin to lose its linear form before a person dies and maybe it can then express itself somehow through the shattered remnants of a ruined brain. Although any such expression seems intuitively to be all but impossible, it's nevertheless quite possibly the *only* way that the phenomenon of terminal lucidity could manifest and is at least not inconsistent with SoS theory. We need to digress a bit and think carefully about the basis of memories in general in order to see whether the 'all but impossible' is quite as impossible as it seems at first hearing.

In the first part of this book, memories were envisaged as 'valleys' in mental landscapes, which was a way of picturing the 'attractors' manifesting in the extraordinarily complex dynamics of mind. The existence of the attractors themselves depends on factors that bias dynamical behaviour. Read most textbooks and you can easily get the impression that the nature of what does the biasing has been pretty well tied up, at least as far as long-term memories are concerned. They are down to Hebbian learning resulting in long-term potentiation in the NMDA sub-type of glutamate dependent synapses, the neurochemical origins of which have been worked out in admirable detail. There's maybe a contribution also from the less well-understood long-term depression, which reduces, instead of increasing, the effectiveness of individual synapses. Well, that's certainly part of the story but can't actually be anything like the full story.

There are a host of memory mechanisms spanning a whole range of clock-time scales. Short- or very short-term ones include gap junction and ion channel opening or closure, reverberating neuron circuits, protein activation (as with CaMKII, for example, described in Chapter 3) and dendritic spine shape changes. Medium- to long-term ones involve gene activation and de-activation, synapse pruning and growth, loss of entire neurons (plus the development of new ones in some areas such as the hippocampus) – and the famous long-term potentiation and long-term depression. Some of the details as to how these mechanisms inter-relate have been worked out, but understanding remains far from complete. Perhaps the main lesson to be learned is that any brain activity of any sort is going to have a clock-time duration and is therefore a 'memory' even if only a very ephemeral one. Some of these ephemera get translated into longer-term format and are then able to mould future behaviour. They become able to generate 'valleys' in the landscape of mind whenever an appropriate dynamical context arises, according to the metaphor I've been using. A proportion of longer-term ones also become amenable to conscious recall, i.e. the so-called 'episodic' and 'declarative' memories that relate to personal history and factual information respectively.

When a long-term memory is consciously recalled it has to be translated back into short-term format. In effect recall involves re-creation of a sort of anaemic (usually) version of the brain state that was originally responsible for the memory. Moreover, if it is to be retained after recall, it must go through at least some of the processes involved in its original establishment. Re-establishment involves opportunities for editing, which is probably responsible for most false memories. If you want to implant a false memory in someone the best way to do it is to get them to rehearse an originally veridical memory as often as possible while introducing subtle, or even not so subtle, pressures to modify the details. People can do this to themselves, of course; it's not always down to interrogators creating the story they want to hear! It's a slower process than inducing false memories under hypnosis, the method mainly responsible in all likelihood for promoting the 'alien abduction' fad (see Chapter 2), but is harder to detect.

The relevance of this from an SoS theory perspective is that consciously recalled memories are referred back to the place where the thread of experience is woven, and get interwoven with ongoing perceptions, cognitions and the like. Is there any evidence, other than that from 'terminal lucidity', to suggest that the thread of conscious mind, woven in what was the past from a clock-time perspective, can directly influence weavings in the clock-time present?

'Evidence' is perhaps too strong a word to apply, but there are a number of phenomena hard to explain on any other basis. For instance, eidetic and 'flashbulb' memories are suspiciously realistic and detailed, as are memories recalled in the course of 'abreactions'. The processes involved in long-term memory and its recall generally lose detail along the way. Where does the extra detail come from that manifests in these types of memory? Maybe some sort of holographic process not involved in 'normal' recall is a possibility that could explain where the extra information comes from. However, it's hard to see how an explanation of that sort could account for the apparently total recreation of past brain states that 'Ruth' manifested, for instance, when her Stroop test performance reverted to its early childhood condition (see Chapter 8). Her case isn't unique in this respect as hypnotic regression can sometimes produce similar, astonishingly complete recreations of much earlier states. The problem for any straightforward neuropsychological explanation is that the brain recreating such states is vastly different in modulator balance, connectivity, neuron numbers and all sorts of other ways from the one originally responsible for the state. It can certainly look as though some influence independent of clock time may contribute to these phenomena.

Then there is the puzzle as to how recall is possible of conscious states dissimilar from ones that a person has *ever* experienced in ordinary life. The classic example was provided by Mozart who said that he could hear his compositions 'all at once' in his mind and then had to go through the sometimes tedious process of writing them down in clock time. How could he possibly have recalled experiencing something that could never have had objective existence (i.e. an entire symphony with all the notes played at the same time)? Similarly, NDEers have talked of experiencing '360-degree vision'. That's not something that could ever have been laid down in their neurophysiological memories from experience of the objective world, while their 'life reviews' are even stranger in a variety of ways. Likewise mystics speak of 'ineffable' experiences that they can't put into words. Nevertheless, though vocabulary may fail them, clearly they remember something of the quality of 'not-of-this-world' experiences. If the experiences are not of *this* world then they must be of the 'subjective' world, in which case it follows that properties exclusive to that world can get incorporated into objective memory states and reports of experience. They can then get rewoven into the ongoing 'thread' of conscious mind.

One could well point out that lots of subjective-only experiences get into memory and can be recalled later, such as ones of flying in dreams or seeing little green men after ingesting psilocybin. Their subjectivity presumably derives from objective neurophysiological happenings induced by sleep or drugs. However, it's arguable that the experiences mentioned in the previous paragraph are different in kind from dreams or hallucinations since they are of qualities that not only don't but also *couldn't* relate to anything existing in the objective world. How could an 'objective' mind bring into consciousness some entity totally unlike anything embodied in its normal 'landscape'? Surely it's at least equally plausible to suppose that back action from subjective mind can influence 'landscape' forms. Perhaps 'terminal lucidity' is merely the tip of an iceberg that happens to show especially clearly an influence of subjective mind on objective mind with its underlying neurophysiology.

The obvious next questions are *where* does subjectivity influence neurophysiology and *how* do effects manifest in physical terms? The general answer is that, since subjectivity is all about meaning and meaning depends on the 'valleys' in the landscape of mind according to the metaphor used throughout this book, while the 'valleys' themselves are formed by memories, subjectivity has to be pictured as affecting early stages of memory processes. But precisely where or how is impossible to say or even guess at present. They are potentially researchable questions, perhaps, but there are others that need to be tackled first. We'll turn to them in the next chapter.

11
NEW DIRECTIONS

I hope you're willing to agree by now that SoS theory, with its basis in a 'landscapes of the mind' picture, is an example of the sort of theory that *might* be useful when it comes to understanding the nature of consciousness. As we've just seen in the last three chapters, it is at least able to encompass some of the more outré phenomena associated with mind that people often like to sweep under the carpet because they elude contemporary neuroscientific understandings. Whether it's a *correct* theory, or offers a suitable basis for elaboration into a correct theory, is an entirely open question. The only other types of theory that are both currently on offer and could in principle compete with it in scope are the 'quantum consciousness' ones such as OrchOR (see Chapter 5). They are, however, relatively hazy when it comes to accounting for *why* conscious experience should be associated with whichever particular quantum phenomenon (usually coherence) they identify as significant, and none of them offers any story that might explain qualia differentiation.

Perhaps the main difference between them and SoS theory is that they don't use the notion of symmetry breakage and don't envisage any deep split between subjective time and the clock time of physics. Like SoS theory they focus on happenings to do with quantum measurements, but they don't take the additional step of envisaging a fundamental divide between consciousness/subjective time and neural eigenstates/clock time; for them, consciousness and neurology both tick along on clock time.

The differences between SoS theory and quantum consciousness ones imply that it should be possible to make predictions of phenomena allowable in SoS theory but forbidden in all quantum consciousness ones – which isn't necessarily an easy task given Richard Feynman's dictum that '*anything* not forbidden in quantum theory is compulsory'. It was a lot simpler to predict previously unknown roles of astrocytes from 'landscape' theory than it is to identify potential phenomena able to differentiate SoS theory from all existing or potential 'quantum consciousness' ones. One might suppose that temporal anomalies could do the trick but unfortunately 'entanglement' considerations, which are atemporal to all appearances, might possibly allow any conceivable anomaly of this sort to be given a 'quantum consciousness' interpretation.

There is a candidate that could do the trick though, albeit one that offers very much of an outside chance. Quantum theory and everything deriving from it is very tightly constrained by energy conservation, which is what its 'Hamiltonians' depend on. As mentioned earlier, energy conservation is, in turn, a consequence of the fact that their physics doesn't change if systems undergo *smooth* temporal transitions. Transitions between the subjective time of SoS theory and clock time cannot, however, be smooth because they are very different entities; the one chunked into subjective 'moments', the other a continuous metric relating to causal interactions. We have already seen that, if SoS entities exist, there is pretty good evidence that they are capable of 'back action' on physical systems, action that must involve a temporal component. There is thus a potential at least for the occurrence of transitions between subjective and objective time that won't be smooth and therefore need not be constrained by energy conservation. A phenomenon that involved violation of energy conservation would thus be incompatible with any quantum theoretical explanation but would be compatible with SoS theory, in that respect at least. Is there any hope of finding energy conservation violating phenomena, and where might one look for them if there is hope?

The point is that *any* violation of energy conservation, even just a single isolated example, would imply the existence of new physics involving a non-smooth displacement in time because such violation is impossible according to thermodynamics, relativity theory and quantum theory. It would be a quite different type of new physics from the varieties sought by string theorists and the like because of its dependence on a temporal discontinuity of some sort. If any such violation also appeared to be dependent on conscious mind, it would provide strong support for SoS theory or some closely related concept.

We've already seen that there is some hope of finding violations since physical phenomena produced in séances, when not down to conscious or unconscious fraud, may possibly have depended on them. However, dusty records of past séances are never going to provide the sort of watertight proof needed to convince anyone of something so remarkable. Stephen Braude's 'gold leaf lady' (Chapter 8) might have rewarded energy balance investigation but she seems to have been unique and any opportunity for that sort of enquiry is probably long gone. There may be more promising avenues to explore than those involving murky invocations of 'spirits' or inexplicable one-offs.

St Simeon Stylites spent thirty-seven years up his succession of pillars, the last of which was said to be around fifty-feet high with a platform at the top about a metre square. He was up there come rain or shine, summer and winter – and winters can be cold in Syria. Presumably he had unusually efficient brown fat to keep him warm since his rags were almost certainly inadequate and there's not much room for callisthenics on top of a pillar. Where did he get the energy he needed? Perhaps he took unrecorded sabbaticals or popped down from his pillar at night when no-one was looking in order to gather food. No doubt his many fans supplied him with goodies although the supply was probably unreliable from time to time, especially during the Persian invasion when he was recorded to have been definitely up there and deterring Persian atrocities with his prayers. There's no knowing nowadays how he maintained his energy balance and survived for so long under conditions that would kill almost all of us within a year. He was far from unique in his endurance, however, as there were lots of stylites and other ascetics scattered throughout the deserts of the Middle East at the time. And there are modern equivalents who might be accessible to investigation.

The modern Simeons are mostly Indian ascetics of various types – yogis, hermits, sadhus, fakirs – who are not necessarily fak*ers*. Every now and again there are reports that one of these has existed without food or, less commonly, water or even air for unfeasibly long periods of time. A recent case who was put under medical observation for two weeks was said to have eaten nothing and drunk only 100 cubic centimetres of water during that period. The latter claim was consistent with an allegedly zero urine output over the whole period. If the claims were true he should have been semi-comatose or worse by the end of the two weeks, but apparently he wasn't. Little effort has been put into investigating claims of this type mainly because they are so self-evidently 'impossible'. Who wants to waste their time looking into what couldn't have happened unless through fakery? If the claims were true, however, broken energy conservation must have played a part in some role or other to prevent the sadhu being poisoned by the products of his own catabolism. Thorough investigation of such cases, with all 'i's dotted and 't's crossed, might prove the most straightforward route to detecting mind-related violations of energy conservation if these ever occur.

Conducting studies of this sort might well be possible for Indian consciousness researchers, of whom there are many these days, since a proportion of sadhus are said to be happy to co-operate with investigation of their capacities. Getting the wider world to believe in the validity of positive findings, if any, would be much more difficult. Maybe SoS theory could help with probable acceptance problems, however, especially if further evidence for its potential explanatory power could be found from other sources.

One of the SoS claims is that there is likely to be two-way interaction between conscious mind and neural memories. Neural memories weave mind but conscious mind can feed its content back into neural memories, according to the theory. And memory must, in any case, be a prime focus of interest for all consciousness researchers because it and experience are so closely interlinked in so many ways. The possibility of reciprocal interactions, and enquiry into their nature, is well worth exploring for its own sake quite apart from any implications for SoS theory.

What the theory adds to more conventional views of any likely reciprocity is a possibility that some of the content of conscious mind may sometimes derive from non-neural sources. It's a little like the Eastern notion of an 'akashic field' in that respect. The principal Western proponent of the idea (Ervin Laszlo, author of over forty books on the topic) first pictured the supposed field as a sort of celestial log book containing a record of everything that has ever happened, which our consciousness can access to a limited extent. He, along with other writers, now regards it as a field of universal consciousness that encompasses 'memories' of everything in so far as it is outside time. Very limited aspects of it manifest in our brains. The analogy usually offered is that the akashic field is like the sum of all television broadcasts and our brains are sets tuned into a particular channel. Perhaps the main problem with the concept is that it's open to so many interpretations that, a bit like psychoanalysis, it's a 'one size fits all' idea. It can be made to correspond with concepts ranging from the 'block universe' of relativity theory to an idea of God as the ground of all being. SoS theory is more focused and thus potentially more useful.

There are several lines of enquiry that could be expected to produce particularly interesting findings, though all would present major technical and procedural challenges. One has to do with the amazing changes in brain functionality that can happen along with hypnotic regression to a 'childhood' self. What are the neural correlates of such regression and where does the information come from that allows such comprehensive resurrection of long-past brain states? A small number of studies have related hypnotic susceptibility to variations in brain connectivity while others have looked at changes in connectivity occurring in the course of hypnotic regression, but it isn't known whether changes in connectivity are functional mimics or recreations of earlier states. And there's no hard information at all as to how these changes are brought about. SoS theory predicts that the conscious mind of the person being regressed makes a direct contribution, not mediated by presently available neural memories, to the genesis of such states especially when they relate to a pre-adulthood past. Realistically, however, testing this prediction is way beyond current technical capabilities, even if CERN levels of funding were magically available. But it may not be so entirely out of reach if and when better calcium imaging methods are developed (or some other method equally capable of directly showing 'mind' in action).

It should be easier to find evidence for an independent effect of conscious mind on neurology when linkage in the brain-to-mind causal direction is weakened or temporarily broken, which brings us back to edge-of-life situations and possibly to psychedelic drug-related phenomena too. The latter may have more in common with edge-of-life happenings than appears on the surface. There's probably no coincidence in the fact that the literal translation from Quechua of ayahuasca (the name for a potent psychedelic brew made from a vine among other ingredients) is corpse-rope! We've already seen that both NDE phenomena and end-of-life ones pose a range of major conundrums to do with consciousness/neurology interactions. How might one go about trying to solve the puzzles?

One straightforward approach to finding out where the information in NDEs comes from has drawn a blank so far. It was inspired by an old report dating back to the 1960s (by psychologist Charles Tart) that a lady who had frequent 'out of body' experiences was on one occasion able to 'see' and accurately report a five-digit number hidden on a shelf high above her bed. Her EEG was being recorded at the time so it was quite impossible for her to have sneaked a look because any large movement would have been very obvious on the recording. People often say that the initial stages of an NDE involve them being 'outside their body' and apparently up near the ceiling. So various items were concealed (from staff as well as patients of course) high up in intensive care units in the hope that an NDEer would spot one and remember having done so. Only a few patients so far have had the right sort of NDE in a bed-space equipped with a concealed object, and none of these reported seeing the object. The numbers are too small to draw any firm conclusion, but it does rather look as though NDE information may not be coming from any simple equivalent of direct visual perception. People do report memories of 'visual' perceptions of environments and objects that are apparently veridical and have sometimes been independently confirmed as veridical, but it may be the case that these recalled 'perceptions' are later constructs deriving from information acquired by other means.

If the information isn't coming from some direct vision equivalent, which was the obvious possibility to test first, there are alternatives to consider that fit rather better than a 'direct vision' hypothesis with experiences reported from later stages of NDEs. After all, it's unlikely that NDEs involve two different types of perception and there's no way that the suburbs of paradise or whatever can own the same sort of 'reality', in ontological terms, as an operating theatre or hospital ward. All conscious perceptions, whether of hospital wards or visions of dead relatives, are subjective constructs and it's conceivable that those occurring during NDEs are informed by 'telepathy' or 'clairvoyance' rather than by some equivalent of an objective sensory input. We have little idea as to what these terms imply, but they refer to capacities with a very small 'effect size' in everyday life. Perhaps the effect size is magnified at the edge of life. Some reported psychedelic drug phenomena are consistent with magnification of these faculties, and the same might apply as life ebbs away. Anyhow, the possibility could in principle be tested. For instance, instead of concealed objects near a potential NDEer, a person holding some strong image in their mind could be used to test for telepathically derived information, or something could be hidden that had emotional salience for the NDEer in a test for clairvoyance.

Testing for telepathically derived information would present fewer problems and might stand a greater chance of turning up positive findings since NDEers generally say that, if they encounter anyone in the course of their NDE, communication is usually 'telepathic' rather than verbal. Perhaps it would be possible to recruit shifts of skilled meditators and position them, while they were holding some randomly chosen image in their minds, near critically ill patients. Would an NDEer pick up on the image? The main problem here is that only around 30 per cent of people who die in intensive care units are successfully resuscitated, and less than 20 per cent of those will report an NDE of any sort. Only about one patient in twenty is going to be in a position to provide the sort of data needed and we don't know what proportion of those, if any, are likely to be 'positives'. A small army of extremely patient meditators would be needed to carry out the experiment, but perhaps it's not entirely impractical.

The problem with testing for clairvoyance is that some object likely to grab an NDEer's attention would be needed; finding an appropriate object amid the flurry and bustle of a crash call would test anyone's ingenuity. According to one report an NDEer successfully spotted where his dentures had been inadvertently hidden, but that was serendipitous and would be hard to formalise in an experimental protocol!

Apart from trying to ascertain where information in NDEs comes from, the other principal question is to do with how such information gets into neural memories so that it can be reported later. If progress could be made with these two puzzles, a host of others could be tackled but these are the two that need to be approached first. The 'how does it get remembered' question can be broken down further and perhaps the most important first step towards dealing with it is to ask *when* the neural memories are laid down. It's very unlikely that they are formed at the apparent time of an NDE because this often appears to coincide with absence of discernible neural activity. As mentioned previously, it's more likely that they are laid down some time after successful resuscitation. SoS theory, with its potential for divorcing subjective temporality from objective clock time, allows for a possibility that what appear from an objective point of view to be memories can continue to exist in the 'present' of a conscious subjective format till a brain is in a state to receive them.

The implication here is that everyone may have these subjective states. Therefore we need to look carefully at people who *don't* report NDEs after resuscitation and compare them with those who do, since the theory suggests that condition at the time of, or soon after, successful resuscitation may be what prevents or allows recall of NDE experience. Maybe oxygen saturation at that later time is what matters, or maybe some other metabolic factor, or maybe medications administered or other factors associated with resuscitation. There are lots of possibilities but their relative importance, if any of them are in fact important, should emerge from the appropriate statistics. And knowing what appeared to *prevent* NDE recall might tell us a lot about *how* NDE memories are established in people able to report them.

The overall message here is that making progress towards understanding conscious mind might benefit from the sort of hands-on proactive approach adopted by our Victorian predecessors, but it's probably better nowadays to focus on rather different questions from those that preoccupied them. Simply gathering data, as they mostly did, is always necessary in the early stages of any field of study. When theory testing becomes practicable, it's better to concentrate on that. They did have an advantage, though, in that séances are a lot easier to conduct than studies of remote sadhus or investigations into apparently haphazard happenings in intensive care wards! Are there any equivalent short-cuts that we could use?

There's the intriguing possibility that psychedelics might prove to be adequate substitutes for brushes with death in some respects and might allow exploration of similar questions to those that can be asked about NDEs. There's lots of interest these days in shamanism. Shamanic 'vision quests' and the like are thought to have often been psychedelic-aided and have quite a bit in common with aspects of NDEs. They too raise questions about where information gained during them comes from and how it gets remembered. The research that was beginning to get under way sixty years ago into a range of psychedelic phenomena was cut short by the 'war' on drugs, which may at last be coming to an end in favour of more rational policies and controls. A few pioneers have resumed formal research (e.g. Rick Strassman's study of DMT effects), but it's still impeded by legal restrictions and funding problems. Perhaps psychedelic research is best regarded as a 'watch this space' possibility, hard to use for legal reasons at present but one that may pay off fairly soon. It would certainly be far easier to determine whether psychic faculties such as telepathy are enhanced by LSD, for example, than to try to ascertain whether NDEers can be shown to have enhanced faculties of the same sort. Some recent anecdotes and pilot studies suggest that psychedelics can indeed enhance 'extrasensory' perception, but rigorous enquiry is still for the future.

There does seem to be quite a strong case for supposing that the most direct route to gaining a better understanding of conscious mind would be through attempting to ascertain where the information recalled in NDEs comes from and how it gets recalled. This would appear to be true regardless of whether SoS theory proves adequate; the theory's main value at this stage is to highlight what may turn out to be the most important questions that need asking. If psychedelics do eventually turn out to be useful substitutes for being on the verge of death when it comes to enquiring into these questions, there's a good chance that answers to them will start to turn up in the not-too-distant future.

12
SPECULATIONS AND IMPLICATIONS

The picture of mind and consciousness developed so far can be regarded as quite optimistic in its way, I suppose. The proposal is that people's essential selves are woven out of tiny chunks of subjective time. These components, dubbed 'SoSs', are envisaged as having an existence independent of objective, clock-time physicality. The 'weave' might well be expected to fall apart when it is no longer attached to its 'loom' (i.e. the brain) with consequent loss of everything that makes a self a self. However, there's pretty good, although far from conclusive, evidence that it doesn't in fact fall apart but may change its overall topology. The old metaphoric image of the psyche emerging butterfly-like from the chrysalis of the body could indeed be apt. We've explored some aspects of the neurology and physics that might endow conscious mind with a form to fit the poetic image. And we've looked at ways in which the theory could be tested and developed – or refuted as the case may be. In the rest of this chapter I'm going to abandon scepticism and will assume that the picture of our nature that has been built up, or something very like it, is indeed true in order to discuss a few of its less obvious social implications.

Obvious implications, should the theory turn out to be true, would of course include a big change in society's attitude to death. Exactly how attitudes would change is unforeseeable, but there's a good chance that the cruelties involved in keeping people alive at all costs, however damaged or demented they might be, would be seen for what they are and mitigated somehow, which would be a great mercy. Presumably too, 'spiritual' values

– the pursuit of truth, beauty and love in the form of *agape* more than *eros* – would move up the social agenda, with also unforeseeable but perhaps mainly beneficial consequences; at least there would probably be long-term benefit whatever the cost in short-term upheaval. Rather than continue to speculate about the unforeseeable, I'd like to take a look at some possible implications of the theory for understanding social history and thus major formative influences on the *content* of consciousness and consequent human behaviour. It's going to be a rather roundabout story.

The story starts with yet another question, perhaps a rather childish one. Why is life so often so ghastly? Some of its nastiness is a necessary consequence of biology. The way the world is set up, we couldn't have evolved without death and all its concomitant sad losses. Distressing consequences of our imperfect genetics – congenital malformations, hereditary diseases and the like – are no more than an inevitable downside of the capacity to evolve. The same can be said of parasitism; freeloading was always bound to evolve as a lifestyle along with symbiosis and commensalism. Competition with other creatures for the wherewithal to live and reproduce is also built into the nature of the world, however distressing its consequences for the losers. We can understand all this and work to minimise these necessary evils.

But humans, as has often been remarked, have added a whole new layer of horrors which rarely serve any obvious evolutionary purpose, not even from a strictly 'selfish gene' point of view. There's no need to greatly labour this point as current events in the Middle East make it well enough. Happenings like those in South Sudan, where peoples freed from decades of oppression and murder by their northern masters almost immediately set about murdering one another, are only too common. Collective madnesses leading to genocides and inevitable destructive backlashes are a recurring feature of history, hard to understand or explain despite the many attempts to do so. Mechanisms enabling or predisposing to such events can sometimes be pinpointed, though generally only with the benefit of hindsight. Nevertheless they almost always leave a feeling of bewilderment. How could perfectly nice, ordinary people behave in the nightmare ways that they so often do behave? Clearly their mental 'landscapes' must undergo some profound transformation to enable them to act as they do. Can we understand anything of the principles behind such transformations?

The fact is that 'cultures' in their many manifestations have, to all appearances, a lot in common with biological organisms. Arnold Toynbee's magnificent *A Study of History* tried to identify a natural history common to entire civilisations from birth to death followed by the growth of offspring. Ours is a 'third-generation' civilisation, he claimed. His work has been much criticised as intellectual fashions have changed but the underlying intuition behind it, that the phenomenology of historical events pertaining to social organisations has quasi-biological characteristics, keeps resurfacing in a variety of forms and contexts. People discuss the 'life' of nations or political parties, for instance, and there is current interest in the 'natural history' of limited companies. The UK Society for Organisational Learning proclaims on its web page that it is 'passionate about cultivating organisations as living systems'. Books describing national or regional 'character' are ubiquitous. And cultures, of course, have huge influence on the mental 'landscapes' of individuals belonging to them. They are emergent properties of individual minds in a manner closely analogous to that in which mental 'landscapes' are emergent properties of individual brain cells. Cultures both emerge from, and feed back to mould, the minds that generate them.

Organisations, even entire civilisations, have hitherto been far less complex than individual brains of course, which may account for many of their crudities. The World Wide Web, however, is beginning to approach brain levels of complexity. The consequences are unforeseeable because feedback between individual minds and an entity so complex is likely to be chaotic. No doubt attractors will eventually emerge to regularise outcomes but how they will manifest is anyone's guess. Some have supposed that a sort of 'super-consciousness' will develop, but that won't happen according to SoS theory since web energies are too dispersed and too incoherent to form the necessary 'weave'. Where SoS theory may prove relevant is in connection with the 'genetics' of cultures.

Richard Dawkins suggested the term 'meme' for genetic units of culture and the idea was taken up by a range of enthusiasts including philosopher Daniel Dennett and psychologist Susan Blackmore. It made a lot of sense in that it is almost as easy to distinguish artefacts and works of art made in China from those of Indian or European origin, for example, as it is to tell at a glance that this particular dog is a Husky, not a Labrador or an Alsatian. Or perhaps I should say that it *used* to be almost as easy to 'see' cultural origins in this way before everything started to homogenise. Nevertheless it is still possible to identify many films, house designs and so forth as being characteristically French, British or Scandinavian. 'Units' of some sort must have got into the different cultures' various products to enable both their commonalities and their characteristic differences.

The problem was that no consensus emerged as to how a 'meme' should be defined or how its basis should be conceived. Was it an idea or something physical? Was it, on the one hand, as extensive as the notion of a car, and was it perhaps something as physical as the actual vehicle sitting in front of someone's house? Should it, on the other hand, refer to the idea or the actuality of something much less complex such as a single windscreen wiper? Clearly memes have to be regarded as contextual. If you're a mum doing the school run, 'car' is probably a single unit for you. If you're a mechanic trying to diagnose a fault, 'car' breaks down into a multitude of separate units. Nobody was ever going to be able to build a cultural 'genetics' from contextual will-o'-the-wisps like these, so it was often argued by anti-meme groups. Robert Aunger tried to resolve the difficulty (in a book titled *The Electric Meme*) by suggesting that memes should be identified with particular patterns of electrical activity in the brain but that didn't settle the contextuality problem, which merely shifted to questions about how some particular pattern might arise. Memes came to be seen by many as an 'explanation' that explained nothing and academic interest in them faded at much the same time that the word 'meme' was being added to English-language dictionaries.

The loss of interest may well have been a mistake because 'memes', if pictured as culturally embedded preconceptions, can be used to account for a range of quasi-medical conditions, such as the 'dancing manias' of the Middle Ages or the various nineteenth- and twentieth-century fatigue syndromes, which elude explanation on other bases. It's possible to trace in some detail the separate preconceptions that coalesced when conditions were favourable to produce outcomes expressed in behaviours or symptoms that were not associated with any of the contributory notions taken in isolation. Indeed the 'alien abduction' experience mentioned in Chapter 2 can be understood as arising from a coalescence of preconceptions, sometimes held subliminally, that had become widespread in twentieth-century America thanks to UFO reports, science-fiction stories and the like.

There are many far grimmer examples of a similar process. For instance, there was a false Roman idea, originally associated with the Catiline conspiracy that Cicero waxed so eloquent about, that secret sects exist who sacrifice babies and drink their blood. The idea kept resurfacing throughout medieval history when it was often attached to anti-Semitic prejudices and helped to inspire pogroms in many regions, especially Germanic ones. It even infected Scottish social workers in the late twentieth century, in only slightly modified form, resulting in a whole batch of children in the Orkney Islands being removed from their homes and taken into care. Similarly the Christian notion of the Apocalypse has probably been responsible, albeit indirectly, for as many deaths in the last 1,500 years as were caused over the same period by some fairly major infectious disease such as measles or typhoid fever. Whatever memes are they can be lethal, although their more usual but less obvious role is a beneficial one in enabling, or in a sense *being*, culture.

Jungian archetypes offer an alternative conceptual approach to these issues. Their manifestations or 'representations' suggest that archetypes share the timeless quality demonstrated by memes. To give a few scattered examples, there is a tribe originating from the margins of the Zambezi floodplain (the Lozi) who could well have found a home from home with the Greeks of Homer's world. Proud, quick, subtle and rather too fond for their own good of fame and possessions, the Lozi have a quite different 'character' from neighbouring tribes despite having a very similar material culture; a character that recognisably manifests the 'hero' archetype. Anthropologists such as Ruth Benedict have described other examples of similar pervasive differences in quality that were not obviously dependent on material circumstances. Then there are the sad manifestations of the 'warlord' archetype that have taken almost identical forms in periods and places ranging from Dark Age Europe to post-colonial Africa – and the later warlord manifestations are not always a case of 'those ignorant of history are doomed to repeat it'. Many of them, and their advisers, have been well aware of relevant history although how they chose to interpret it is another matter. At a more sophisticated level, it's remarkable how similar were the legal and social arrangements for coping with lepers in thirteenth-century England to those for looking after mentally ill people in the early twentieth century. The associated public prejudices and other attitudes were similar too, as were even the proportion of asylum places made available per head of population.

Perhaps one might suppose that it would be better to think of these memic and archetypal phenomena in terms of 'extended mind' – with its embodiment in shared stories, cultural practices and artefacts – being moulded by 'laws' of sociology and psychology, with the outcomes feeding back to affect the content of individual minds. In fact, however, thinking that way would merely be a more cumbersome way of approaching the issues since memes and archetypes are shorthand terms *for* 'laws' of sociology and psychology, in the sense that they refer to factors that channel social and psychological outcomes rather as Ohm's law can be said to channel the behaviour of electrical circuits. 'Memes' refer more to small-scale, local 'laws' and 'archetypes' to larger-scale ones, though there is a lot of overlap. Both are terms that describe attractors in the landscape of social dynamics. Seen this way, there is no longer any mystery about the contextuality of memes since the configurations adopted by 'landscapes' of this sort are always context dependent.

As we saw at the start of this book, attractors (i.e. the 'valleys' in dynamic 'landscapes') are memories, which implies that memes and archetypes must be viewed as manifestations of memories. They are embodied in oral traditions, writings, customary ways of doing things, artefacts of all sorts. In line with Dawkins' original proposal, memes, like genes, are essentially packages of information that help to shape individual minds but which endure and are reproduced down the generations. They provide a 'genetics' for our cultures and social organisations just as genes themselves provide information that helps to shape our bodies and brains. And a few of them, as he said, can behave like his 'selfish genes' causing mayhem and misery.

SoS theory adds additional twists to the tale. Because memes and archetypes enter conscious minds they get represented in 'subjective' format as well as in the 'objective' customs, artefacts and the like that Dawkins envisaged. Indeed, since they are basically ideas, their primary form *is* subjective and their objective representations secondary. Since we are, in a very real sense, the content of our subjectivity, there is thus a sense in which we are, to a quite large extent, the memes and archetypes that we harbour. This is equivalent to saying that our minds and personalities are moulded by the societies in which we grow up, but perhaps expresses the point more forcefully.

With their principal home in 'subjectivity', it's also the case that these entities must be pictured as occupying subjective time, not clock time. The implications are very unclear, but there's certainly a possibility that they might gain an appearance of trans-temporality or extra-temporality from an objective, clock-time perspective. They might be supposed to indeed possess the sort of numinosity that Carl Jung attributed to his archetypes. People are often happy to think that the saints of this world may be pure, if partial, expressions of Christ or Buddha archetypes, while regarding the Hitlers as being more like flotsam thrown to the top of a heap by some social process or group mentality run amok. But maybe both great saints and great sinners alike are embodiments of trans-personal archetypes and their associated memes, manifesting in unusually undiluted form.

Both saints and major sinners have generally been known for their charisma; a quality that most of us recognise but that is difficult to define. It's not what Freud referred to as 'ego-inflation' – that occupational hazard for celebrities, professors, CEOs and politicians – which often, but not always, goes along with charisma. Indeed some saints with great charisma are said to have lacked any hint of ego-inflation. It's a rare quality, and recognisable only through personal contact. I've encountered only four people myself who seemed definitely to possess it. One was a psychiatrist (William Sargant), another an astronomer and TV presenter (Patrick Moore), the third a pioneer asset stripper (Eliot Slater) and the fourth a president (Kenneth Kaunda). They were a mixed bunch, but each managed somehow to convey a special quality. These people represented what archetypes may be like when expressed in unusually pure form at some stage of a person's life. They owned a sort of numinosity, which was probably quite independent of any moral value in what they expressed.

The relevance of all this is that social outcomes and behaviours may be moulded not only by the objective chains of causation that historians and others try to identify but also by subjective iterations of particular varieties of selfhood. Christian saints have sometimes affirmed that 'it is not me, but Christ within me', and maybe they were expressing a truth less metaphorical than is often supposed. The trouble is that Stalin might perhaps have claimed with equal truth that 'it is not me, but Ivan the Terrible within me'. Improving human behaviour may therefore be even more difficult than social reformers usually suppose due to a form of trans-personal, trans-temporal inertia in the subjective world that is liable to generate selves predisposed to be awful along with the relatively few predisposed to be especially not awful. The process can be envisaged as feeding back to bolster institutions in the objective world more likely to perpetuate horrors than bring benefits. It can be regarded as yet another inevitable, if remote, consequence of constraints consequent on the requirements of evolution, which is often liable to favour the generation of individual subjective selves more predisposed to vice than virtue.

Luckily biology and other factors prevent most of us from expressing archetypes in any pure form. Our selves are generally made up of a mixture of memes from a medley of sources, in so far as they derive at all from cultural sources. Our biological predispositions must often get in the way of expression of any subjectively embodied memes. Someone well-endowed with mirror neurons and thus empathy, for example, is never going to whole-heartedly accept another of the memes of Roman origin that still persists and causes mischief: namely the idea that torture is a good way of ascertaining truth. Looking at social history overall, one can envisage that there must be a sort of randomisation process going on with some memes or archetypes happening to fall on stony biological ground and others finding congenial homes.

It's a process that would have much the same function in relation to the social 'landscape' as does sleep in relation to landscapes of mind; a process of shaking things up, so to speak, to prevent 'valleys' from becoming too deep or too extensive. A purely 'subjective' world might well be unable to host any similar function, in which case it would tend to crystallise out into rigid forms. The interplay between subjective and objective worlds may be thought to maintain 'flexibility' in the former at the expense of some increase in the 'rigidity' of the latter.

Is that the purpose of our existence in this vale of tears – to provide a form of 'sleep' that maintains the liveliness of subjectivity? If so, the horrors of the world are not purposelessly cruel as they often seem. They are nightmares inseparable from the delights and beauties of our existence, which are also the dreams of the subjective world. The stuff of our objective life, with all its pains and all its joys, enables vitality and variety in the subjective world – or so at least SoS theory can be taken to imply – and perhaps that's a function sufficiently worthwhile to justify the collateral harms that are all too apparent. Maybe SoS theory retains an underlying optimism despite its implication that redeeming human behaviour is likely to prove a task far beyond the capabilities of any simple programme of social reform.

13
LOOSE ENDS AND NEW BEGINNINGS

The picture that's been painted in this book shows conscious mind as woven from threads of subjective time into a tapestry that depicts patterns of energy manifestation in the brain. The objective world, built from chains of causation, includes and extends into our brains where it contributes to dynamic 'landscapes' that channel and shape brain energies. The picture depends on two principal conjectures: first that the 'time' figuring in the Heisenberg time/energy uncertainty relationship is indeed a subjective time that differs from the clock time implicit in objective causal processes. The second conjecture is the supposition, which accords with everyday experience, that meaningful patterns of energetic events in brains retain their meanings when translated into a subjective format. It was suggested in this connection (in Chapter 7) that knot theory might provide a suitable concept of the translational principle needed to allow meaning inherent in patterns of brain activity to correspond to that in qualia.

If these two conjectures aren't wildly astray – and the second at least fits common-sense – we've got a pretty good outline understanding of how neural activity informs conscious mind. It's quite heavily dependent on various metaphors and mathematical abstractions, but I hope these have been illuminating rather than misleading. The details of processes that they represent are inordinately complex. Abstractions are needed for any understanding of what may be going on, in much the same way that 'temperature' is more useful for most purposes than thinking about the kinetics of individual molecules. What we haven't got is any good understanding whatsoever of how conscious mind informs neural activity, although there is evidence from a wide range of sources that 'back action' of this sort can and does occur. All that's been offered are some rather vague remarks about early stages of neural memory processes being *where* consciousness must mainly affect our brains, which is true enough in all likelihood, but doesn't tell us anything about the *how*. That's surely the biggest loose end of the whole theory.

There's good evidence from NDEs in particular that consciousness *does* affect neural memories. Then there's pretty good evidence from a range of sources that these effects are not necessarily mediated by any relativistic causal mechanism. Finally there's the evidence, also pretty good, that minds can affect physical objects beyond the brain in ways inexplicable in terms of any known principles of objective causation, while conscious intentions can weakly bias outcomes of probabilistic events occurring in the laboratory. It's presumably sensible, according to Ockham's razor at least (i.e. 'don't multiply entities beyond necessity'), to look for a common basis on which to explain these rather disparate phenomena.

I had thought, until very recently, that there probably weren't any ideas around that could offer the least hope of allowing the sort of detailed account of what might be going on that one would like. I'd thought that the best one could do was to make rather vague, general remarks of the sort offered at the end of Chapter 8, where it was said in effect that 'it's something to do with conscious mind's special temporality'. This was useful in the Chapter 8 context especially because of its implication that violations of energy conservation may be possible, but doesn't provide any sort of basis for understanding how detailed memories could translate from a 'subjective' into a neural format; how, in other words, the 'back action' of consciousness on neurology might be mediated. Then a 2015 paper by physicist Maurice Goodman, in an online journal that specialises in out-of-the-box thinking (*Journal of Consciousness Exploration and Research*, vol. 6, no. 1), made me wonder whether it might be possible to do better.

Goodman argues that the weak nuclear force may have some special role in quantum biology and consciousness. At first sight the idea looks like a total non-starter since the force itself operates only over intra-atomic nucleus distances, not even molecular distances (because its vector bosons are so massive), while neutrinos arising from the radioactive decays that it mediates hardly ever interact with matter – you need a huge detector to 'see' even one or two of the zillions that zoom through the detector every minute, most of which come from the sun. However, contemporary detectors respond to high-energy neutrinos only (the higher the energy the more sensitive they are) and it could be the case, for all we know, that very low-energy ones have wholly different interaction characteristics with matter, rather as low-energy (infra-red) photons warm us while high-energy ones (X-rays) mostly go straight through us.

Goodman goes on to point out that the weak force, unlike the other three forces of nature, has the right energetic and other characteristics for any quantum coherent phenomena in which it might be involved to operate over biologically significant distances and consciousness-relevant clock-time durations. That could be coincidence, of course, and doesn't go far towards mitigating the range and interaction problems when it comes to proposing a role in biology. However, there's a further 'coincidence' in that the delocalised 'size' required of electron neutrinos, if they are indeed to have a biological function at the cellular scale, predicts a mass for them. The actual mass is currently unknown for sure but should be pinned down more precisely within a year or two by the Karlsruhe Tritium Neutrino Experiment (KATRIN). Goodman's prediction for the mass (of ~ 0.16 eV/c^2) falls well within the expected range (which is 0.05 eV/c^2 to 0.5 eV/c^2). If proved correct the prediction would lend credence to other aspects of his theory, since two 'coincidences' such as these must make one wonder whether they were in fact coincidental, and would be a very significant achievement in itself, perhaps at least as significant in the long term as was prediction of the existence of the Higgs boson.

From the point of view of SoS theory, there are also pointers to a possible role for the weak force since it's the only asymmetrical force of the four. It is 'left handed' while the other three are 'ambidextrous', suggesting that it is connected with a broken symmetry. This isn't the electro-weak symmetry break mentioned in Chapter 1, by the way, since the electromagnetic force has no preference for 'right-handedness'. Could there be some connection with SoS theory's subjective/objective break? Two hints suggest that there might. First, the weak force's asymmetry (broken 'parity') has a connection with the direction of time's arrow because overall charge/parity/time symmetry ('CPT' symmetry) is preserved. It's a connection that isn't well understood, like almost everything else to do with time, but could be taken to suggest a link between the weak force and SoS theory's subjective time/clock time disjunction. Second, the weak force is responsible for radioactive decay, clock-time rates of which were shown to be affected by conscious intention, albeit to a very small extent, in the PEAR experiments.

It looks, in brief, as though it might be conceivable that a broken subjectivity/objectivity symmetry ties in somehow with the weak force, while coherent phenomena involving the force may prove capable of affecting biological functions on the scale of entire cells or more. It's a possibility that encompasses many ends so loose that they are more than likely to fly off in the wind when tested, but at least it suggests that an explanation of conscious mind's effects on neurology may be possible one day. At present, though, it's a case of 'watch this space'.

Goodman's suggestion makes a range of interesting and potentially testable predictions, apart from that of electron neutrino mass; for instance, that low-energy neutrinos exist and are able to interact readily with matter to affect biologically important functions. If it can be linked to SoS theory, there would be an implication that the broken parity of the weak force is another consequence of the subjective/objective split thought to occur along with energy eigenstate 'measurements'. In which case, pre-measurement weak force states should presumably still be 'ambidextrous'. Questions as to whether low-energy neutrinos interact with matter in a different way from high-energy ones and when the weak force loses parity symmetry are probably answerable in principle at least, but might never have been asked in the ordinary course of events. Whether they are *worth* asking remains unclear of course, but at least they open up possible avenues of enquiry that have an outside chance of paying off, however small it may be.

And that's really what this book has been all about; it represents an attempt to expand the 'enquiry space' of consciousness studies in potentially fruitful directions. I think it's fair to say that mainstream enquiries in the late twentieth century were hobbled by constraints that are only now beginning to lose their grip. Philosophers, on the one hand, were often preoccupied with purely semantic issues that could never be resolved without better understanding of the concepts to which their terms referred. They were in much the same situation as were early nineteenth-century thinkers discussing 'energy' prior to the development of thermodynamics. Neuroscientists, on the other hand, were shackled by a dream – or nightmare some might say – inherited from the Enlightenment: the image of ourselves as computerised automatons. Even now there are billion-euro plans to simulate brains on computers. It's safe to say that they will tell us something about which aspects of brain function *can* be simulated but they won't tell us much about mind and nothing about the nature of consciousness, for the model that inspired the plans has such very limited validity.

We need models that relate to the actual phenomena out there in the world and in our minds, not to dreams of, or preconceptions about, what should be there. Taking the phenomena seriously was something that Victorians such as William James, Frederic Myers and many others were seriously good at. Their problem was that they lacked any adequate concept of materiality, though their knowledge of neurology and psychology was good enough for their purposes, while they were constrained by ambient religious preconceptions prevalent in their culture just as much as are many modern neuroscientists by anti-religious preconceptions. Their research programme faltered and failed for those reasons, not because of any intrinsic misdirection. Re-adopting their approach now to broaden enquiry offers a way of breaking out of constraints that are stifling progress. Luckily many researchers are beginning to take the phenomena seriously again; most notably NDEs, which, unlike some other types of phenomenon, have achieved a degree of cultural acceptability thanks to the innumerable accounts of them that have been circulating for the last forty years.

The ideas described in previous chapters have pointed to a range of lines of enquiry that may prove to be especially fruitful, and it's worth giving a brief reminder of them. Some are mainly to do with neuroscience, others with physics and the rest with anomalous phenomena, though there's quite a lot of overlap between the three categories. I'll sort them into those that involve relatively conventional approaches and the rest:

First, the conventional ones:

- Looking at the dynamics of calcium-ion behaviour in the context of the 'landscape' picture of mind. Are the dynamics fractal or pseudo-fractal, spatially and temporally? Can regularities (i.e. 'attractors') be identified that relate to particular percepts or cognitions?
- Continuing enquiry into how general anaesthetics work. SoS theory suggests it would be worth trying to ascertain whether reduction in total brain energy usage is critical and/or whether anaesthetic agents increase the energy uncertainties involved in energetic events that are very precisely 'measured' in their absence.
- Comparing the immediately post-resuscitation states and variables of people reporting NDEs with those not reporting NDEs in order to try to ascertain which variables, if any, may influence recall.

Next, some unconventional ones:

- Given the somewhat way-out suggestion that knots might be relevant to qualia, is there any evidence that calcium-ion dynamics are ever braided and/or manifest Seifert surfaces?

- Can evidence be found that conscious mind is involved at any stage in producing 'pre-sponses' (see Chapter 4), or is the phenomenon entirely a correlate of unconscious mind or biology? Experiments making use of 'masking' techniques might go some way to providing an answer.
- Do psychedelics enhance parapsychological effect sizes in the same sort of way, perhaps, that Ganzfield techniques have been said to improve remote-viewing success rates? There are anecdotes and pilot studies suggesting that psychedelics may have this property, but no rigorous studies (yet).
- Can hard evidence be unearthed for the occurrence of failures of energy conservation connected with the activity of conscious minds?
- Can NDEers report the unspoken thoughts of people present while they were being resuscitated? There are lots of anecdotal reports that they can and have done so, but these need to be formally tested.

The suggestions about calcium-ion dynamics are probably going to have to await better imaging methods before they can be properly tested, but the others are all both doable and overlap with the interests of some present-day researchers. Luckily a range of social trends suggests that public interest in the whole field is growing, with the prospect that they might be properly funded in the near future. Reports in the media of anomalous phenomena are less dismissive and debunking on the whole than was the case twenty years ago, though still too liable to confuse 'alien abduction' or 'I saw a UFO' type phenomena with the more genuinely puzzling ones. But the really big change is probably being driven by concerns over the ever-increasing numbers of damaged and sometimes demented elderly people and the unanswered questions about how best to care for them and how we should think about their deaths. Research into issues related to such questions is surely overdue and badly needed from the points of view of both individuals trying to think about their situation and that of society trying to do its best for the elderly along with others in life-threatening situations.

A tentative conclusion reached from looking at the evidence so far available is that we are nothing much like the computerised automatons, inherited from the Enlightenment, of current mainstream techno-culture. Rather we are perhaps, as Prospero averred in *The Tempest*, 'such stuff as dreams are made on'. It's even conceivable that our existence in this 'objective' world is to a larger 'subjective' world as sleep is to us within the 'objective' world. In that case, Shakespeare may have been wrong to make his character conclude that 'our little life is rounded with a sleep'. In relation to the 'subjective' world, it could be the case that our little objective life *is* a form of sleep. Trying to make a start on finding out whether there may be some truth in such a view, or whether it is just another blind alley, is surely more than a little worthwhile.

APPENDIX

Synopsis of the arguments

Chapter 1

Two separate threads of opinion about human nature persisted into the twentieth century. One, inherited from the Enlightenment, viewed people as biological automatons. The other affirmed that they possess 'spirits'.

Late nineteenth-century investigations into our nature derived from earlier interest in mesmerism and spiritualism.

Late twentieth-century, culturally 'mainstream' enquiries into consciousness centred on analytic philosophy, neuroscience and cognitive psychology. However, developments in fundamental physics in particular, along with philosophical arguments, led to proposals that conscious experience and the 'objective' world are ontologically distinct in some way.

Although phrased in different language and employed in different contexts, it appears that concepts underlying 'mainstream' and 'spiritualistic' pictures of consciousness share commonalities.

Chapter 2

The content of our minds (both conscious and unconscious) can usefully be pictured as the content of dynamic state spaces, deriving from both brain and body/environmental dynamics (perhaps in a proportion of around 9:1). The principal advantages of this picture are that it

- allows explanation of group and extended mind phenomena;
- accounts for our need for sleep;
- shows how genetic, personal and socio-cultural memories are at the basis of mentality;
- shows how we are able to assign meaning to information so quickly and in a manner that digital computers are generally unable to match;
- accounts for some of the downsides of our type of mentality, including 'change blindness'.

Chapter 3

Shows how the type of mind pictured in Chapter 2 is likely to depend on fractal patterns of electrochemical activity in brains. Calcium-ion dynamics are centrally important to this activity, particularly because of their close links with early stages of memory processes many of which involve enzyme (CaMKII) activation.

Because of the requirement for fractality, it can be predicted that astrocytes must have important parts to play in 'mental' functioning (in addition to the 'brain housekeeping' roles traditionally assigned to them). Evidence that they do play such parts is mounting rapidly.

Neurotransmitter and neuromodulator complexity is likely to have evolved in order to allow rapid, large-scale adjustments of mental 'landscapes' to suit changing physiological or environmental conditions.

Chapter 4

Describes how the time measured in physics ('clock time') refers to successions of local causal interactions, whereas the time that we experience is of sequences of 'present moments' each with extended clock-time duration. 'Now' is a meaningful concept in experiential time, but is of strictly local relevance only in physics time.

The direction of time's arrow (pointing from past to future) is normally common to both physics and experiential time, due to entropy on macroscopic scales and the irreversibility of quantum 'measurements' on microscopic ones. However, there is a now well-established research finding (variously named the 'pre-sponse', 'presentiment' or 'feeling the future') showing that present physiological and other happenings can foreshadow future events to a very limited extent (i.e. with a small 'effect size'). It is not currently known whether future *conscious* experience has a role to play in producing this anomaly.

Chapter 5

Discusses currently popular philosophical, neuroscientific and physics-based views of the nature of consciousness. Surveys some of the evidence on which these views are based, together with limitations of both evidence and theories. Generally, philosophical arguments, on the one hand, are excessively constrained by semantics; neuroscientific ones, on the other hand, must invoke philosophical 'dual aspect theory', 'property dualism' or similar notions in order to account for phenomenal experience, while physics-based views struggle to cope with problems arising from 'decoherence'.

It is pointed out that the 'landscape' view of mentality described in Chapter 2 is functionally similar to two of the most popular neuroscience-based theories, namely 'global workspace' theory and 'integrated information' theory. However, it has greater explanatory power in several respects than either of these older theories.

Chapter 6

It is assumed that neutral monism in the form of the Pauli/Jung conjecture correctly describes the basis of reality. There are indications that the broken symmetry giving rise to our separate subjective and objective worlds occurs along with energy eigenstate manifestations. It then follows from the Heisenberg time/energy uncertainty relationship that time is primarily subjective and energy manifestations primarily spatial. Each energy measurement made at any given location will thus be associated with an irreducible chunk of time, dubbed a 'scintilla of subjectivity' (SoS) and having a definite, clock-time, objective duration. Most energy measurements have large uncertainties and therefore almost infinitesimal SoSs, which implies that their 'subjective' components can be ignored for practical purposes, rather as the 'virtual particles' that are spatially ubiquitous can be ignored.

Patterns of energy manifestation in brains, especially any with minimal energy uncertainty, will map to patterns made up of overlapping chunks of subjective time (SoSs). This picture suggests that the 'binding problem' of neuroscience is actually a pseudo-problem. It also has at least as much in common with the 'spiritualistic' concept of consciousness identified in Chapter 1 as with the mainstream 'new orthodox' view.

Chapter 7

An attempt to identify concepts capable of accounting for how one quale can differ from another – why isn't all our experience simply undifferentiated 'subjectivity'? The idea of a 'qualia space' with dimensions contributed by all the different factors associated with qualia looks promising at first sight, but leaves questions about how such a space could support the apparently irreducible distinctiveness of different qualia. Knot theory, however, offers entities equally 'at home' in spatiality or temporality that are also irreducibly distinct from one another in the case of prime knots. Knots are likely to occur in brains and knot theoretic considerations could thus bridge the gap between objective and subjective worlds while preserving irreducible distinctiveness. The theory is therefore an example of the *sort* of bridging concept that is needed, whether or not it eventually turns out to be the correct concept.

It transpires that there are mathematical conjectures linking prime knots to prime numbers and even to energy eigenstates, suggesting that there are indeed ill-understood links between brain energy eigenstates, knot theory, prime numbers and, by extension, qualia. Carl Jung's remark that his 'archetypes' are 'like' the natural numbers, which inspired this line of thought, may thus turn out to have been prescient.

Chapter 8

A look at anomalous phenomena that especially interested late nineteenth-century investigators in order to decide which particular ones might be most likely to provide clear evidence relating to SoS theory. Hypnosis and telepathy are considered unlikely to give readily interpretable evidence. Apparitions, although startling and unusual, are probably not anomalous except when they convey veridical information unknown to a perceiver – in which case they can be regarded as telepathic phenomena and thus also unlikely to provide useful evidence.

Anomalous physical phenomena that most readily lend themselves to investigation fall into two categories: first, those manifesting during séances; and second, the very small biases in probabilistic happenings demonstrated in the PEAR experiments. The former are often thought to be 'impossible' due to their apparent violation of energy conservation. However, energy conservation relates only to smooth, clock-time, temporal displacements (via Noether's theorem). SoS theory, with its disjunction between clock time and experiential time, allows that non-smooth temporal displacements may be possible, in which case energy conservation might be violated. The PEAR experimental outcomes might be envisaged as consequences of a mental 'landscape', manifesting in SoS format, that extends to include pre-measurement aspects of the experimental apparatus.

Chapter 9

SoS theory views conscious mind as 'woven' by brains into patterns from threads of subjective time. Does the weave fall apart when 'its' brain does so, or can it appear to persist from a 'clock-time' perspective? Evidence from children who 'remember' previous lives strongly suggests that it doesn't necessarily fall apart; that from 'past-life regression' is much weaker albeit pointing to the same conclusion. There is also surprisingly strong evidence from mediumistic channelling that not only do individual woven patterns persist but also that they can change and develop from our clock-time perspective.

Chapter 10

SoS patterns are capable of 'back action' on mental and neural functioning, so the evidence suggests. NDE reports and the phenomenon of 'terminal lucidity' offer the best prospects for exploring the nature of any such action because they appear to involve complete or near-complete reversal of the normal brain activity → conscious experience flow of causation. Some reports can be interpreted as suggesting that SoS pattern topology may switch from its usual linear state to a more 'spherical' condition during an NDE. Back action must primarily influence neural memory processes in some way. Issues to do with memory are discussed and potentially researchable questions identified.

Chapter 11

A discussion of potentially fruitful lines of research. Unambiguous demonstration of violation of energy conservation associated with conscious mind would distinguish SoS theory from any conceivable 'quantum consciousness' theory; suggestions are made about where it could be looked for.

Research into NDE experience might most usefully be directed towards trying to ascertain whether the veridical information sometimes reported has 'telepathic' origins and enquiry into what distinguishes people who don't report NDEs from those who do. SoS theory suggests that immediately post-resuscitation variables are likely to prove the most relevant.

There's a possibility that psychedelic drugs could sometimes provide adequate substitutes for brushes with death in relation to some of these enquiries.

Chapter 12

A look at implications of the views arrived at for understanding social history. It is suggested that the trans-temporal nature of SoS patterns adds an extra layer to the deterministic processes that can loosely be pictured as 'memic' in origin. Social reform is thus likely to prove more elusive than many hope. However, there's a final speculation that the turmoils of this world may serve to ameliorate any tendency to excessive rigidity in the subjective world; our objective world may thus have a function in relation to the subjective one equivalent to that of sleep in relation to our objective selves.

Chapter 13

A brief statement of lines of research that should prove worth pursuing in the future.

INDEX

AI (artificial intelligence) see 'computers'
akashic field 111
alien abduction 19, 20–21, 94
Alzheimer's disease 4, 104
anaesthetics, general 57, 128
Apollo (temple of) 1
apparitions 7, 85–87
archetypes 78, 91–93, 120–121
astrocytes 32–34
attention 59
attractors 23, 25, 61, 120–121
automata 7, 12
ayahuasca see 'psychedelics'

Baars, Bernard see 'global workspace'
back action 57, 75
Balliol College (master of) 3
Barbour, Julian 44
Bateson, Gregory 26
behaviourism 8, 21
Bem, Daryl 46
binding problem 50, 55, 56, 70
Bose–Einstein condensation
  see 'OrchOR theory'

Cajal, Ramon y 53
CaMKll 32
causation 10, 42
Chalmers, David 49

change blindness 24
charisma 121–122
clairvoyance 97–98, 103, 113
coherence, quantum 62
coloured dot illusion 45–46
combinatorial problem 26
complexity overkill (of
  neuromodulators) 36–37
computation, quantum 27
computers, computation 12, 23–24, 26
consciousness 2, 38, 49–64, 74; see also
  'SoS'
consciousness, quantum theory and 61–64
conservation laws 66, 89, 84, 109
counterfactuals, quantum 10
cross correspondences 96
cultures 117–120
culture wars 1, 8

Darwin, Charles 8
decoherence 50, 51
delayed choice experiments 10, 43
Descartes, René 4, 11, 16, 70
Diana, Princess 19–20, 21
DMT (dimethyltryptamine) see
  'psychedelics'
dualism, property, aspect 14, 15–16,
  50–51
dynamic state spaces see 'landscapes'

EEG (electro-encephalography), electromagnetic fields 30–31, 34, 55–56, 85
electromagnetism 16
enchanted loom see 'Sherrington, Charles'
end-of-life experience 103–104
energy 67, 68, 71, 76, 83–85, 97, 129
entanglement, quantum 10
entropy 34

facial recognition 20
fanaticism 38
Faraday, Michael 7
feeling the future see 'pre-sponse'
Fields, Douglas 34
fMRI (functional magnetic resonance imaging) 34, 54
fractals 29–30, 31–32, 34, 35
Freeman, Walter 30
Freud, Sigmund 7, 11

Galileo 73–75
galvanic skin response (GSR) 46
gap junctions 27, 42
gauge theory 53
global workspace 60, 61
gold leaf lady 87, 110
googling 3
Grey Mary 52

Hameroff, Stuart see 'OrchOR'
Hamiltonians 80, 109
Heim, Albert 92
Heisenberg uncertainty 67, 69, 94, 124
hippocampus 31, 56
holography 35
HOT (higher order thought) theory 52
human nature 1–2
hypnosis, hypnotism 7, 18, 84–85
hypnotic regression 94, 95, 112

information 26, 60–61, 112–113
ions, calcium 30–31, 128

Jackson, Hughlings 53
jaguars 59
James, William 45
Journal of Consciousness Studies 4, 52
Jung, Carl 50–51, 52, 78

karma 94
ketamine see 'psychedelics' and 'anaesthetics'
knot theory 78–80

Lakatos, Imre 83
'landscapes' 22–23, 27–28, 36, 38, 47, 59–60, 61, 71, 122–123
Laplace, Pierre-Simon 5, 9
Leibniz, Gottfried 11
Libet, Benjamin 52
localisation, functional in brains 54–55, 57

materialism, materialists 11, 13–14, 49
materialism, promissory 14
meaning 25–26, 107
measurement, quantum 9, 42, 68
meditators 113–114
mediumship 95–97, 103
memes 118–121
memory 23, 27, 41, 51–52, 78–82, 85, 102, 104–106, 111, 114
Mesmer, Anton 6
mesmerism 6–7, 12
Mettrie, Julien de la 5
mind 2, 18–28, 38, 69–70
mind, embodied, extended 27–28, 120
mind, holism of 21–22
mirror neurons 37
Mormons 18–19, 20
Myers, Frederic 12, 84

Nagel, Thomas 74
NDE (near-death experience) 92, 99–103, 113–114, 128
neural activity, inhibitory 44
neural nets, artificial 27
neural nets, small world 53
neuromodulation 30
neurons, single 31, 56–57
neutral monism 50–51, 65
neutrinos 126, 127
Newton, Isaac 5, 10, 40
Noether's theorem 66, 89

oligodendrocytes 34
OrchOR (orchestrated objective reduction) theory 62–63
orthodoxy, scientific old and new 14–15, 56

Index

panpsychism 50
parapsychology 9, 14, 84
parity 126
Pauli, Wolfgang 50–51, 80
PEAR (Princeton Engineering Anomalies Research) experiments 89–91
Penrose, Sir Roger *see* 'OrchOR'
Pereira, Alfredo Jr 34
photons 16, 36
physical anomalies 68–69
pineal gland 4–5, 16, 50
poltergeists 7, 87
present, specious 45
present moment 41
pre-sponse, presentiment 46–48, 129
Pribram, Karl 35
psychedelics 57–59, 112, 115
psychokinesis 87–90

qualia 49, 72, 73–81
qualia space 77, 81
quantum counterfactuals 10
quantum theory 10, 61, 109

reality 82–83
recognition, facial 23–24
reincarnation 93–94, 95
relativity theory 9, 10, 68; *see also* 'time, clock'
res cogitans/res extensa 4, 5, 8, 9–10, 14, 15, 16, 70, 72
research directions 97–98
Reynolds, Pam 100
Riemann hypothesis 79, 80
Russell, Bertrand 50, 51
'Ruth', story of 85–87

Sapir–Whorf hypothesis 75
Scientific and Medical Network 14
séances 88–89
Shannon, Claude 26
Shanon, Benny 58
Sherrington, Charles 11–12, 38, 81
single neuron recording 45

sleep 24–25, 31, 93, 98
small world networks 42–43
Smolin, Lee 44
Society for Psychical Research (SPR) 8, 14, 84
SoS (scintillae of subjectivity) 68–70, 72, 78, 60, 72, 75, 79, 80, 93, 96
soul 4, 5
spirits 4, 6, 12–13
spiritualism 8–9, 12–13, 14, 15–16
'split brain' findings 55
St Augustine 40, 44
St Simeon Stylites 110
Stargate programme *see* 'clairvoyance'
Swedenborg, Emanuel 6
symmetry 15, 66–67, 83, 96, 126
synaesthaesia 76
synchrony, gamma 45, 67

telepathy 85, 87, 113, 129
terminal lucidity *see* 'end-of-life experience'
tesseracts 29
time, arrow of 42–43, 71
time, clock 40–42, 54–55, 68; *see also* 'energy'
time, experiential 44–46, 68–69, 72
TMS (transcranial magnetic stimulation) 54
Tononi, Giulio 60, 77
tool use 24, 28

universe, clockwork 7
unus mundus 65
Wallace, Alfred Russell 9
weak nuclear force 16, 125–126
Wheeler, John 10
Windbridge institute 95
Winter, Alison 8

Xavier, Chico 95

Z particle 16
Zeno effect, quantum 43

For Product Safety Concerns and Information please contact our EU representative GPSR@taylorandfrancis.com
Taylor & Francis Verlag GmbH, Kaufingerstraße 24, 80331 München, Germany

www.ingramcontent.com/pod-product-compliance
Lightning Source LLC
Chambersburg PA
CBHW070622300426
44113CB00010B/1626